WE BE HERE WHEN THE MORNING COMES

WE BE HERE WHEN THE MORNING COMES

Foreword by
Robert Coles

Text by **Bryan Woolley**
Photographs by **Ford Reid**

The University Press of Kentucky

Design by Jonathan Greene

ISBN: 0–8131–1337–7

Library of Congress Catalog Card Number: 75–18285

A statewide cooperative scholarly publishing agency serving Berea College, Centre College of Kentucky, Eastern Kentucky University, Georgetown College, Kentucky Historical Society, Kentucky State University, Morehead State University, Murray State University, Northern Kentucky State College, Transylvania University, University of Kentucky, University of Louisville, and Western Kentucky University.

Editorial and Sales Offices: Lexington, Kentucky 40506

We dedicate this book
to our friends
LOUIE & RUBY STACY
and all who go
into the pits

CONTENTS

FOREWORD

ALTHOUGH the region called Appalachia is often described as isolated, removed from the American mainstream, there is a substantial body of "literature" devoted to the people of Eastern Kentucky and West Virginia. Not all of these articles and books are worth much. Many of them, for decades, have stressed the ignorance, the backwardness, the deficits and limitations of the region's people—those "hillbillies" whom the rest of America left behind decades ago. As for coal miners, their only distinction is that they are employed hillbillies, in contrast to those who distill liquor illegally, who speak, at best, a quaint English, and who come forth occasionally with "interesting" arts and crafts as examples of their "folk culture." I need not elaborate further on the uninformed condescension of it all. Much of it comes from scholars who claim to have done long and thoughtful "studies" of people—people who come across, in the write-ups of those studies, as peculiar, strange, hopelessly different from the rest of their countrymen.

How convenient for the rest of us, including those who own and run the region's major industry, the coal mines. If the people of Clover Fork, Kentucky, some of whom we meet and get to know in this important and powerful book, could be portrayed convincingly as seriously retarded culturally, as not very smart, as provincial and even a little odd, then who is to worry about the way the rich and influential corporations treat them? Aren't they lucky just to have jobs? And aren't they, really, limited enough in mind and spirit to justify (for those always trying to come up with excuses) the way they live? Could these people really "adjust" to the better life enjoyed by those of us who read (and of course write) books? Would they even want "our" life?

Are they not, after all, happy in their own "culture"? Or, if not happy, then inextricably wedded to it—to the "culture of poverty," which some social scientists tell us persists over the generations, a modern secular equivalent, of sorts, of the concept of "original sin," a common and inescapable heritage?

I hope many social scientists, among others, will read this wonderfully strong, passionate, candid, and, yes, scholarly book—scholarly in the sense that the people who speak in it know whereof they speak (about themselves) and do so with intelligence, perception, thoughtfulness. They are short on theory, no doubt about it. They speak plain, understandable English, not a mumbo-jumbo social science dialect full of cloudy phrases and terms. And they are inclined to be direct, unashamedly emotional, forceful, and, not least, analytical about their ideas and feelings. I only wish some of us in the social sciences, some of us who call ourselves sociologists or political scientists or psychiatrists, would be as blunt and as plain spoken in the way we come to terms with who owns what in this country, who profits from whom, and who pays what price to get what wage while others rake in an avalanche of profits.

While going through this book I was seized by dozens of memories—memories of the quite similar people I have met over the years as I have worked in western North Carolina, Eastern Kentucky, and West Virginia. It is all too easy today for someone like me to become yet another American "expert"—one who claims to know so much when in fact he may have only begun to learn a little. Over the years much of what I once believed (and, alas, wrote about) has been proven incorrect by those who ought to know, the people I have been visiting and talking with. Once I saw Appalachia's people as sadly "disadvantaged," as "culturally deprived," as fitting all the rest of that litany of words, phrases, and thinly disguised pejorative labels that people like me are wont to bestow on them. If I know a little better now, it is because of particular men, women, and children up particular hollows, who have patiently and with kindness insisted that I be set straight.

People like them, maybe even kin of theirs, are here in the pages of this book, brought to us by a good and persevering writer and a sensitive photographer. I only hope a lot of us in this rich and powerful but by no means altogether "just" nation will be set straight by the impressively honest and honorable people who speak through this book.

ROBERT COLES

PREFACE

THE LOT of the Appalachian coal miner is easier now than it was forty years ago. Boys get to grow up before they go underground now. There is no starvation in the camps. Men are paid by the day and not by the carload, and they are paid well. They no longer owe their souls to the company store. There are few company stores left now. Mechanization has made the work less brutal. Federal and state legislation and unionization have made it safer and more secure economically.

But the history of the Appalachian coal industry is a sorry tale, full of broken promises, blasted hopes, lies, bitterness, violence, and death. Few improvements in the miner's lot have been bought cheaply, and not all of the leaders he has followed have been heroes. Nor have the blessings won by the strikes, battles, protests, and negotiations of the past benefitted all who go underground. Work and living conditions, dreams and expectations vary widely from county to county, hollow to hollow, mine to mine.

We Be Here When the Morning Comes is not a sociological study of "the Appalachian coal miner today"; nor is it a history of the United Mine Workers of America or labor trouble in Harlan County, Kentucky; nor is it an objective report on the Brookside strike of 1973–1974. It is intended to be simply a portrait, in words and pictures, of a community and its people at a particular instant in their lives. It is about life at the bottom of a very tall corporate totem pole, as seen by the people who live it along the Clover Fork of the Cumberland River. It is based almost entirely on our own observations of persons and events and many hours of tape-recorded interviews and conversations with miners, miners' wives, and veterans of Harlan County's several decades of labor strife. The people are expressing themselves during a period of extraordinary tension and conflict in their lives. It is also about the struggle of a proud, hard-working people for dignity and a measure of self-determination. Since most of the story is told by those people themselves, it is not unbiased.

When Ford Reid and I conceived the idea of this book, the miners of Brookside, Kentucky, had been on strike for a year. Brookside and its sister mine, Highsplint, are owned by Duke Power Company of Charlotte, North Carolina, the sixth largest utility company in the United States and one of the nation's largest coal-burners. The men of Brookside and Highsplint, and others like them, see the nation's life blood—electric power—in its primitive state, while it is still buried in the mountains. Reid and I decided to go there, live with a miner for a while, and describe his life.

We considered ourselves experienced Eastern Kentucky hands. During our years on the staff of *The Courier-Journal* and *The Louisville Times* we both have covered many stories there, and Reid had already photographed some episodes of the strike. But we did not really know what we were getting into.

We drove to Harlan and discovered that no one wanted to talk to us, much less let us live with him. The international representatives of the United Mine Workers of America, from whom we had expected help and encouragement, were skeptical and evasive. A striker whose name we had gotten from a lawyer friend agreed to several meetings, but stood us up every time. Everybody said we had come at the wrong time. Arnold Miller was about to bring thousands of out-of-state miners into Harlan to march through the streets of the town, and both community leaders and union people were predicting violence. In the string of tiny communities along Clover Fork, where the Brookside mine is, everyone was reluctant to talk. The striker who allowed a writer and a photographer to live in his house would make himself and his family targets in a war which they expected to erupt at any moment. That was the consensus.

It did not change after Miller's parade came and went without a violent incident, for two days later a striker was shotgunned by a company man. We gave up.

I had a second purpose in Harlan. I had been assigned to interview a local attorney, Eugene Goss, for a story entirely unrelated to the Brookside strike or coal mining. I would complete that assignment, then Reid and I would return to Louisville. During the course of the afternoon we spent together, I told Goss about our project and our failure.

"I think I can help you," he said. He telephoned John Deaton, Jr. (called "Junior" by everyone), the proprietor of a grocery store at Ages, Kentucky, around a curve from Brookside. Junior is a former coal miner, and his brother Darrell was one of the strikers. Junior said that he and Darrell would help. The next day I moved in with Louie Stacy, a roof-bolter at the Brookside mine. Reid moved in later, and we remained until a week after the strike ended.

Obviously, this book could not have come about without the help of a great many people, some of whom we would like to thank here:

Our wives, who had to keep explaining our long absences to small children; our editors, Geoffrey Vincent and Al Allen, who also must have wondered about us from time to time; Robert Deitz and Dr. Samuel W. Thomas, who gave us the initial opportunity; Karl Forester and Eugene Goss, who offered aid and comfort when things looked bleakest; Mary Lawrence Young and Anne MacKinnon, invaluable researchers; Betty Mainord, who transcribed hour after difficult hour of tape-recorded conversation with amazing dedication and enthusiasm; and Phyllis Rutherford, for her careful and conscientious typing of the final draft.

We owe a special debt of gratitude to the people along Clover Fork who opened their homes, lives, and memories to us during a difficult time.

Bryan Woolley
Louisville, May 4, 1975

The operators made a union man out of me, the way they treated me. They made me what I am, and I'm glad of it.

The Reverend Otis King

You know, an old dog, if you keep kicking him, first thing you know, he'll turn around and bite you.

Betty Eldridge

CLOVER FORK,
AUGUST 1974

1 CLOVER FORK

Harlan County lies like a short, fat worm along the southeastern border of Kentucky, nestling the Virginia line. It is a high, rugged place in the Cumberland Plateau of the Appalachians, almost isolated from the rest of the world. The broad interstate highways do not touch it. The only railroad is a spur line that wriggles its way up the creeks from Pineville. The Louisville & Nashville Railroad's big locomotives roar in with their long strings of gondola cars, get their loads of coal, turn around, and roar out again.

Cumberland Gap, where Daniel Boone led settlers through these hills, is not far beyond the county line. For generations the people of Harlan County led the life of those first pioneers, getting their living from the fruits of the great forest, and its game, and the little crops they managed to scratch out of the steep hillsides and the bottomlands of the shallow creeks. Since the early days of the twentieth century, though, Harlan County has lived on coal. In the county seat, called Harlan Town by the old ones, stand two monuments to coal. One, a squatty shaft in the middle of a busy highway, is a hazard to traffic. The other, a sort of wall across one corner of the courthouse lawn, bears a plaque commemorating the departure of the first carload of coal from the county on August 23, 1911. Both are constructed of blocks of coal, mortared together. They are monuments to themselves.

When news trickles out of Harlan County to the rest of the world, the news is about coal. When the miners are docile and keep filling up the gondola cars, there is no news from Harlan County. When there is trouble in the coalfields, and the stockpiles of the big industrial plants in the North and East begin to dwindle, the editors of newspapers and magazines and the news directors of television networks pull out their maps and look for

Harlan County. Coal operators there are known to be muleheaded and ruthless and slow to learn. The miners there are known to be stubborn and quick-tempered and violent. Forty years ago the placed earned a nickname—Bloody Harlan. Whenever labor trouble erupts, it is a good bet that Harlan will be bloody again.

In July 1973, labor trouble came to Brookside, a middle-sized mine in Harlan County employing about one hundred and eighty men. The mine is operated by Eastover Mining Company, which is owned by the huge Duke Power Company of Charlotte, North Carolina. The miners at Brookside voted to join the United Mine Workers of America, the union whose organizing efforts—and the operators' violent reaction to them—gave the county its fearsome name back in the 1930s. Contract negotiations between the union and Eastover quickly broke down. Both sides settled in for a long strike. There were occasional picket-line confrontations and courtroom quarrels between union men and company men during the ensuing year, but there was no threat of an old-fashioned, all-out coalfield war until July 1974, when the Brookside men succeeded in shutting down Eastover's Highsplint mine, nine miles farther up Clover Fork.

The fork is a shallow, dirty stream that begins on Black Mountain, the highest point in Kentucky, which is also its border with Virginia. It tumbles between steep mountainsides toward Harlan Town, where it merges with Poor Fork to become the Cumberland River. The L&N tracks and Kentucky 38, a narrow, winding highway, accompany it much of the way. Driving eastward from Harlan Town's black quarter, always uphill or down and always around a curve hugging the steep mountainside, first on one side of the fork and then on the other, one can see only a sliver of sky between the two high ridges. Huddled along the highway, in an almost unbroken line from Harlan to Highsplint, thirteen miles up the fork, are nearly a score of communities that teemed with coal miners and their families a half-century ago but have since shriveled. Many mines have been worked out. Some small operations became hopeless in the face of more stringent federal mine safety regulations in 1969. The remaining mines depend on machines and electricity now, and do not hire as much human muscle and sweat as they used to. But the communities, each with its roadside store and its gas pumps, have held on to their names, and some of them, their post offices.

So you wind by Clovertown, and Kitts, and Golden Ash, and Blackjoe, and Coxton. You cross a narrow iron bridge over the fork and you are in front of a squat, ugly red brick building with a black-and-yellow sign over the porch:

EASTOVER MINING CO.
Brookside, Ky.

Behind the building, on the mountainside, is the driftmouth of the mine and its big sheet-iron tipple. Below the building, along the narrow bottom of Clover Fork, are the few remaining rows of tiny, company-owned frame houses, their privies, and the faucets where the miners' wives go with buckets to get water. Three hundred families used to live at Brookside, they say. But the mine works less than two hundred men now, and in August 1974 only ten or twelve families remained in the camp. Some had been evicted during the strike, and the company had torn down their houses. Others had just found better places to live, around the curve at Ages, perhaps, or farther up the fork at Verda or Draper or Evarts, or along one of the little roads that branch off the highway and wind up even narrower hollows to Jones Creek or Redbud or Kenvir or Bailey's Creek.

Nearly everyone who lives along Clover Fork was born there, or arrived a long time ago from some other mountain county in Kentucky or Virginia or Tennessee. Nearly everyone is kin to someone else there, and they have much in common. They are not highly educated. Some finished high school. Many did not. Large families are still more common there than in the cities. The religious among them are some brand of Baptist or Pentecostal. They cherish a well-stocked pantry above every other material comfort. They grow much of their food in backyard gardens and take pride in setting an abundant table for company. Nearly all of them have a console color TV and a stereo and a large stack of country music, usually on eight-track tapes. The more prosperous or prudent among them live in comfortable frame houses with carpeted floors and paneled walls, or in new house trailers. Those with lesser jobs or large families who cannot afford such homes see that the old coal-camp houses are kept spotless, inside, at least.

All go into the mines, or get their living from those who do. This fact binds them together. It also divides them. Some who live there hate the United Mine Workers of America. Others would die for it.

2 BUCKSHOT IN THE HEAD

LAWRENCE DEAN JONES and Billy Carroll Bruner had been friends. Billy, at thirty-nine, was sixteen years older than Lawrence, and he was a foreman. Lawrence was one of the younger miners at Brookside, but they had liked each other, and had had some fun together. They were neighbors on Jones Creek, and had known each other for a long time.

When the strike came, they wound up on opposite sides, but there did not seem to be any ill feeling. As a company man, Billy was expected to cross the picket line. No one bothered him. After mining ceased at Brookside, the company gave him a bossing job at Highsplint. When the Brookside men began picketing Highsplint in force, Billy was one of the company men who loaded his van with strikebreakers and tried to run them past the pickets. Although Lawrence was a staunch UMWA man and served his allotted time on the picket line, he was not one of the more active strikers. He had married after the strike started, and he and Diane, his sixteen-year-old bride, were now the parents of a three-month-old daughter, Amy. He spent as much time as he could with them. He took a part-time job at a sawmill to supplement the one hundred dollars a week strike benefit he received from the union. He also did odd jobs from time to time.

On August 24, 1974, a sweltering Saturday, he delivered a load of hay to the Reverend Homer Jackson, one of his neighbors on Jones Creek. As he was driving back down the hollow, he saw some friends sitting on a log and stopped to talk to them. He told them he had encountered Billy earlier in the day, and that Billy had pulled a gun on him and threatened him. Lawrence thought Billy had been drinking. Although he did not take the foreman's threat too seriously, Lawrence was a little worried, his friends said later. This is what they said happened next:

While Lawrence was talking with his friends, Billy drove up in his truck. One of the friends got up and approached the foreman. "Ain't nobody here wants no trouble, Billy," he said. "Nobody's got anything against you."

Billy knocked the man out of the way with the butt of his shotgun and pointed at Lawrence. "There's the monkey I want," he said. Then he fired. Lawrence was hit in the head and fell. His friends drew pistols and fired at Billy as he retreated to his truck. Billy was hit, but was able to make his way home.

Lawrence's friends drove him to the Harlan hospital. Kentucky State Police quickly arrested Billy and charged him with assault with intent to kill. He was taken to the hospital for treatment of his wounds. An angry crowd of miners and miners' wives soon gathered there. Billy's wounds were not serious, and as soon as he could be released, the State Police spirited him away to the Bell County Jail at Pineville.

Sunday morning, Deaton's Grocery at Ages was buzzing with the news. While blond little girls in church dresses skipped in smiling to buy bubble gum and lollipops and giggle at Junior Deaton's teasing, their fathers stood angry at the counter, revolver handles protruding from their hip pockets and their waistbands.

"I heared Lawrence was dead before they even got him to the hospital. Is that right?"

"I heared they put machines on him. Pacemakers or something. He's as good as dead, though."

Junior Deaton, a short, chunky, thoughtful man, sat on a high stool behind the high counter, raking in the children's pennies and nickels, making change for the bread and milk their parents bought, plucking down cigarettes from the rack on the wall by the cash register.

"I heared Bruner was killed, too. Dead on arrival."

"Naw, he ain't dead. The state polices taken him to jail in Pineville."

"Too bad they didn't take him to jail in Harlan. He'd be dead by now."

"Yeh, buddy."

Junior accepted several coins from a pretty teen-age girl with a ponytail. "How's your granddaddy, honey?" he asked.

"Just barely hanging on. We're praying for a miracle."

"I hope you get one, honey."

She smiled. "We're happy, though. Granddaddy was saved last night. He made his confession to the preacher."

Junior watched the pretty girl slam the screen door and run across the highway. "I don't believe I knew Lawrence," he said.

"Sure you did. He was in here a lot. Had a little goatee. Real quiet feller. Never gave nobody no trouble."

"I just can't place him."

An old pensioner, hunched and skinny, sidled up to the counter and peered at Junior from under drooping lids. "What do you hear, Junior?" he asked softly.

"I been hearing all sorts of things. Everybody's got a different story."

"Wonder what *really* happened."

"We'll never know."

"Damn gun thugs! I seed too much of this stuff back in the thirties. Reckon it's gonna be like that all over again?"

Junior shook his head slowly. "Sure hope not."

"You and me, buddy."

A miner came in to pay for some Texaco he had pumped into his pickup. "It says three dollars, Junior."

"Yeh, but the price is twice what it shows on that gauge now. Energy crisis, you know."

"Yeh. Awful, ain't it? Say, I heard Lawrence and that Bruner was kin. Is that right?"

"Don't know. They both lived up there on Jones Creek."

Sunday traffic is always heavy at Deaton's Grocery. This Sunday, though, the men lingered longer than usual, with their Dr. Pepper and RC Cola, sharing their anger, pondering the effects of buckshot on a human head.

"We oughta go and get Bruner out of that jail."

"Yeh."

"Well, if he ever sets foot in Harlan County again, he's dead."

"Yeh, if he ain't went, he better get gone, and not never come back."

"What do you hear, Junior?"

"All sorts of things. Everybody's got a different story. I don't pass them on."

"I heared Bruner accused Lawrence of putting tacks in his driveway."

"I heared that."

"Who would shoot a man over a thing like that?"

Another man chuckled bitterly. "A scab'll do anything, buddy."

The slow, angry drawl continued into the bright, humid afternoon. A stocky, sandy-haired man leaned on the counter and asked Junior the price of twelve-gauge shotgun shells. Junior told him.

"You better give me five," the man said. He was Louie Stacy.

3 LOUIE

LOUIE AND RUBY STACY were reared in coal camps. They reared their five children in coal camps. A few years ago they bought a little piece of land down in the Clover Fork bottom between Coxton and Brookside, put a two-bedroom house trailer on it, and moved off company property. Their home and their small yard were surrounded by thick borders of zinnias and marigolds, where bees buzzed, monarch butterflies flitted lazily, and a hummingbird darted in and out among the red and gold petals. Inside the trailer, Cricket, a black Chihuahua dog, barked at strangers. Outside, a dozen nameless cats of various sizes and colors preened themselves on the front steps and stalked field mice through the weedy remains of the garden.

In the small kitchen one humid afternoon, Ruby poured a cup of hot, black coffee and set it before her husband. He lit a cigarette and said:

Franklin D. Roosevelt was the only real president a working man ever had. No doubt about that. I remember the last of Hoover's days. I was real small, but I remember walking seven or eight miles to Pennington Gap, in Virginia, and getting the Red Cross stuff they used to give out for the people to live on. I remember back when we was on starvation.

I was born in 1925. I was about seven years old. But even being that small, it's burnt right into my brain. I can remember the day just as good as if it had happened yesterday. We had to walk all them miles to get that Red Cross stuff to live on. Because back then, my father, he couldn't find no work. And then, when Roosevelt went in, he started this WPA for all the men, and the three C's—we called them CC camps—

and the NYA for young boys to work at. So it give everybody a little bit of income.

I went into the mines when I was thirteen years old. It was a little truck mines over in Virginia, where I grew up, above St. Charles. That was back when you wore a carbide light. We didn't have hard-shell hats then. We had a soft-shelled hat. You know, the top of it was cloth, but you had a leather bill on it, and then you had a hook to put them carbide lights on. It was in 1938. That's back, you know, before World War II. How come I got started, my father was sickly, and he wasn't able to work. He got badly ruptured in the mines, and he worried a lot, too. He had a nervous breakdown, and he had this large family to support, so I quit school and went to work.

Well, the guy I got the job from, he knowed I was way too young, but he agreed to give me a job just because he's sorry for me, you see. And when the inspector would come around, why, he'd tell me to go off in some old works. You know, working in the mines, you got works way off miles back in there, where they done already got the coal out. He'd say, "You go off in them old works, and when the inspector gets here, you blow out your light and hide." So I'd get back in there and hide, you know, until the inspector left. The darkest dark you ever saw, back in there. I mean, there's nothing darker than the inside of a coal mine with your light blowed out.

That was back when there wasn't no such thing as a continuous miner or a motor to pull coal with or a shelly-car or anything like that. That was back when we used ponies in the mines. Pulled coal with ponies. I used to drive one little old pony that had to have a carbide light hisself. He wouldn't go nowhere without that light on his head. When the light started getting low, he'd slow down, and when it went out, he'd stop. I'd get the light to going again, and he'd work fine.

And there wasn't no coal-cutting machines to cut coal with. You had to drill it with a breast auger. You'd drill your coal with just pure old muscle power. Then you'd put your powder in the holes, and shoot the coal down, and load it, and pull it out with ponies. Buddy, in them days, if you didn't grow muscles on top of muscles, you'd die in the mines. They'd kill you.

And that was when you had these little old screens, and you'd dump the coal out, and you'd have to take a shovel and push it over the screen, you know, and your real fine coal fell out on the ground. You couldn't sell it back then. They throwed it away. Only thing you could sell was the big lumps of coal, you know. Sold it for house coal. Trucks would come in out of Tennessee and places like that and haul it away. They throwed that dust and stuff away. And when World War II broke out, well, them hollers over there was full of stoker coal—the little stuff—and dust. But when the war broke out, there was a good sale for it, so they cleaned up all them hollers—every bit of that stoker and dust—and sold it.

And then, after I got a little bit older, I went to Bonnie Blue mines, over in Virginia. It was a big mines. I guess they worked a thousand men or more. I was eighteen then. That was my first big mines. Bonnie Blue belonged to Blue Diamond Coal Company. And I worked there for a while. I guess I worked with a dozen different mines over in Virginia during them years. I believe I was working at Monarch when Ruby and me got married. Got married in 1945, and in 1946, we had a bouncing baby boy, four days before our first anniversary. Ruby had them fast, like popcorn popping, for several years after that.

About five years after the war was over, the work got awful bad over there, so I moved over here, and I been here in this area—from here to about four or five miles on the other side of Harlan—for the last twenty-four years. With the exception of the time I had to leave up here in '64. I was gone five and a half years, because I got starved out, you know.

See, I worked at Brookside in '64, when they had the United Mine Workers up there. They'd had it for years and years. And when our contract was up in '64, we come out on strike. That was under Tony Boyle's administration, and they was giving us a twenty-five-dollar-a-week store voucher for strike benefit at the time. That's all they was giving us to live on. And I stayed on strike up here for a year. We come out in June of '64, and I stayed out till June of '65.

At that time, I had five kids, and was trying to keep them in school, and they all got naked and barefooted, and, well, you can imagine what it was, keeping five kids on twenty-five dollars a week. To keep them in school and keep them in clothes. We was eating potato soup and powdered milk and powdered eggs from that commodities program the government used to have, and was wearing almost nothing. The soles of the kids' shoes was flapping. Kids laughed at them at school. It was pitiful. We was living at the Brookside camp then, and the mines was a Harlan Collieries mines. It belonged to the Whitfields. Old man Whitfield. I don't remember his first name. Don't know as I ever heard it. At that time, we had a county sheriff that had about two hundred deputies, and they was all against us. We had the state polices, and they was against us.

I stayed on the picket line a year. Most of the miners working at Brookside then was older fellers, just waiting around for their pension. And the company had done hired in an entirely new bunch of scabs. I mean, they had a full crew, and they's working six days in the week. And we'd stand out there on the picket line, and they would go by and laugh at us. Actually laugh at us. After a year, I couldn't take it no more. We was starved out, and I saw we wasn't gonna get no contract from the Whitfields. So I took off and went to Detroit. There was still a few old fellers coming out to the picket line, but the strike was really over.

I lay all that stuff on old Tony Boyle. That son of a bitch. I wish them people hadn't got to him in time to pump that bunch of pills he took out of his stomach. He'd been better off dead. I mean, the way he let us down. I'd say if he'd of stayed in the United Mine Workers—if he'd been the president of it—I'd give him five more years, and it'd been broke. He just wasn't no good. He just faded away and left us, there *in '64. The union ain't been at Brookside since then, until now.*

But we've got some good boys in the union now. They're young, and tough, and they ain't gonna give up. And we've got us a good man at the top, too. When we went to Washington, you know, to talk to them congressmans? Why, after it was over, we went to his office. Arnold Miller's office. We sat there and talked for two or three hours. He's a coal miner hisself, and he knows what a coal miner has to go through to make a living. And he told us that we had one hundred and twenty thousand coal miners behind us, and they had sixty million dollars, and every coal miner and every dollar of that money was behind us here at Brookside. And I believe that, too.

I believe he's going to be as good as John L. Lewis. At one time around here in this part of the country, when you said something about John L. Lewis, you'd make sure you said something nice. Cause if you didn't, you'd wind up over in a ditch somewhere. And I believe Arnold Miller's going to be just as good as John L. Lewis, if not better. And this pension plan for these old-timers, you know, that's retired? They get a hundred and fifty dollars a month now. And he says that if he can get this thing organized down in here by the 1977 contract, he's going to try to get them retirement pensions up to five hundred dollars. So that's going to be a good thing. Along about then, I'm going to be thinking about retiring, too. See, I'm forty-nine years old now.

4 ON THE HIGHSPLINT PICKET LINE

THE UNION meeting that Sunday afternoon at the Evarts Multi-purpose Center was tense. The big meeting room was hot and stuffy, and miners looked longingly out the windows at children splashing in the swimming pool that Great Society money had built. Many of the men who had stopped by Deaton's Grocery earlier in the day were there, and perhaps a hundred more. Their tempers were still smouldering. Mickey Messer, the crew-cut, bearish president of the infant Brookside local, confiscated a reporter's tape recorder and placed it on the table beside him. "We don't want no recordings of our conversation today," he said.

It was a special meeting, called to ponder the shooting of Lawrence Jones, to vent grief and anger, and to think of what to do next. "I thought this was 1974, not 1931," Messer told his union brothers, "but it looks like I was wrong."

A retired miner rose and counseled an eye for an eye. "We've been shot at and beat up and arrested," he said, "and we ain't done nothing. It's time we commenced to do something. There's lots of trees and bushes on these hills alongside this road. They'll hide a lot of men. We oughta get some good men with high-powered rifles up there, and pick off these gun thugs when they drive by. There wouldn't be nothing to it atall."

The union organizers rose and pleaded for patience. That's just the kind of thing the company wants to happen, they said. They're just waiting for somebody to lose patience and blow up the tipple. Then it will all be over. No, don't be provoked. The picket line is the place to show your colors. Show up at High-splint in the morning, they said, and bring a carload of friends with you.

In the pre-dawn morning, Louie hunched over his kitchen table. He sipped at his strong black coffee, then poured it into another big cup on the table, sipped again, then poured it back into the first cup. "Coffee cools faster that way," he said, and he hated to be late to the picket line.

He glanced at the calendar on the wall by the telephone. "What's today? August the twenty-sixth? This damn strike's thirteen months old today." He spoke quietly, for Ruby was still asleep, and Cricket was still curled up in his basket in the bathroom. Louie got up and filled his cup again, and lit another cigarette.

He appears younger than his forty-nine years. He is a strong-looking man, handsome in a rough-cut, frontier way, although his face is sometimes slightly distorted by the large dose of snuff he tucks in his cheek when he cannot smoke. He likes snuff better than the chewing tobacco that most of the miners use. It is not so sweet, and it better satisfies his craving for tobacco down in the mine, where the men are forbidden by law to smoke. Louie runs a roof-bolt machine five miles inside the mountain. His job is to keep the mountain from caving in, and he likes it. He would rather do it than sit under the trees and watch men whittle, as he had done almost every day since the strike began.

He left the cups one-third full on the table and put on a tan sweater. He lifted his single-shot twelve-gauge from its corner by the door, opened the breach, and removed the shell he had placed there before he went to bed. From the top of the television he picked up the five shells he had bought the day before at Deaton's Grocery and stuffed them into his pocket. He switched out the light and stepped into the darkness toward the old Volkswagen parked just beyond the border of marigolds and zinnias. With care he laid the shotgun across the back seat.

The engine groaned like an old dog rudely awakened, then coughed into life. Louie flicked on the headlights, and then the windshield wipers to clear the dew. "If the company's dumb enough to try to work them men today, there'll be killing," he said. "It'll turn into a blood war."

A few hundred yards eastward on Kentucky 38 the little car

crossed the narrow iron bridge over Clover Fork. Lights burned in a few of the company houses along the river bottom, and behind the broken, wire-latticed windows of the brick building with the sign over the porch. Eastover's company store had been in that building, before the strike shut it down. "This is where the women beat up them scabs," Louie said, pointing through the car window.

Several miners' wives, including Ruby, had gone to jail for their actions during the strike. Some had taken their children with them. It had been their first time behind bars, and it had been a frequent topic of conversation on Clover Fork ever since. The retreat of the strikebreakers before the angry, switch-wielding women had ended Eastover's attempts to work the Brookside mine, and had entered the labor-trouble folklore of Clover Fork and Bloody Harlan.

Louie's car sped up the narrow, winding highway, past Deaton's Grocery and the tiny sheet-metal Ages post office, through the formerly company-owned villages of Verda and Kildav and Draper, around a curve near the Multipurpose Center where a miner and three company-paid sheriff's deputies were killed in a labor-trouble shootout in 1931, to Mack's Supermarket in the center of Evarts. The parking lot and roadside were crowded with cars, trucks, and campers, some practically covered with UMWA bumper stickers. Men moved among the vehicles, appearing briefly in headlight beams, then disappearing into the darkness again.

Suddenly the vehicles, almost one hundred of them, filled with armed and angry men, began moving onto the highway again, this time in convoy. They rolled quickly through Coates to the village of Shields, turned off the highway, crossed the L&N tracks, then pulled up in line along a narrow dirt road that ran parallel to the highway, but slightly above it, along the hillside. The parked cars formed a rampart between the highway and the miners. Behind the men was a small corn patch and then the steep mountainside. Its thick second-growth timber and underbrush were slowly greening in the dawn's gray first light. Beyond the rooftops of the village, the still black ridges loomed above their fog-laden hollows like the humps of a gargantuan monster rising out of a misty sea.

Louie pointed. "Highsplint's about a mile farther up the road. We used to set up our picket line there, but there was no cover for us. We moved it down here, where we could have some protection when they open fire on us."

The National Labor Relations Board had ruled that the Brookside men had the right to engage in lawful picketing at the Highsplint mine, also owned by Eastover and Duke Power. Production stopped there July 8, and on the next night, Minard Turner, a retired miner and moonshiner, was shot twice in the chest as he approached a car to ask the driver to honor the picket line. The driver sped away, leaving the old man lying in the roadway. The shooting signaled the beginning of the long period of tension and violence that had led now to the shooting of Lawrence Jones. Rumors had spread along Clover Fork that the company had bought machine guns. Many miners and miners' wives claimed to have seen them, and some said they had been fired upon with them.

Governor Wendell Ford had sent dozens of state troopers to Highsplint from outside Harlan County to escort workers past the picket line into the mine. The presence of the troopers seemed to increase the possibility of violence, and after a conference in Frankfort with Arnold Miller, international president of the UMWA, the governor agreed to withdraw them and entrust the peace to the State Police regularly assigned to Harlan County.

"Them state polices was sent in here to break the strike," Louie said, "and they seemed to have a personal grudge against us. The only thing we ever did to them was heckle them, call them Ford's storm troopers and stuff like that. But they strutted around there like roosters and slung them sticks and gritted their teeth. You could tell they had a personal grudge."

The men, some still yawning and rubbing sleep-swollen eyes, crawled out of the cars into the morning and milled into small clusters, exchanging off-color jokes, inquiring about the success of garden crops, bumming cigarettes and coffee from those who had remembered to bring them. Every miner was armed: young men with revolvers holstered and tied to their legs, gunslinger fashion; grizzled, emaciated pensioners with old .45 automatics tucked into their waistbands; squirrel-hunters with shotguns cradled in their arms.

"Hey, buddy, that's a real oldie you got there!"

"Yeh, my daddy carried this back in the thirties. It's a scab-getter from way back."

"Well, it's likely to get another chance today."

Two State Police cars pulled off the highway and parked just down the hillside from the miners. "I hope them boys knows enough to get out of the way when the scabs come," an ancient pensioner said. "I wouldn't want to hurt no state polices."

"I heared that the federal mediator told Duke not to try to work Highsplint till Thursday."

"If they got any sense, they won't try at all."

"Yeh, buddy, but they ain't got no sense."

After a while, the police cars pulled away and disappeared over the hill toward Harlan. The miners remained until seven o'clock, smoking, talking, peering down the highway toward the same hill where the company convoy would appear, if it was to appear. Then, in small groups, they got back into their cars and drove slowly away.

Ruby prepared a breakfast of eggs, biscuits, gravy, sausage, and coffee. Louie ate, then fell asleep on the sofa.

5 SCABS & GUN THUGS

JACK LONDON wrote, many years ago:

> *After God had finished the rattlesnake, the toad, the vampire, He had some awful substance left with which He made a scab.*
>
> *A scab is a two-legged animal with a corkscrew soul, a water-logged brain, a combination backbone of jelly and glue. Where others have hearts, he carries a tumor of rotten principles.*
>
> *When a scab comes down the street, men turn their backs and angels weep in Heaven, and the Devil shuts the gates of Hell to keep him out.*
>
> *No man has a right to scab so long as there is a pool of water to drown his carcass in, or a rope long enough to hang his body with. Judas Iscariot was a gentleman compared with a scab. For betraying his master, he had character enough to hang himself. A scab has not. Esau sold his birthright for a mess of potage. Judas Iscariot sold his Saviour for thirty pieces of silver. Benedict Arnold sold his country for a promise of a commission in the British army. The modern strikebreaker sells his birthright, his country, his wife, his children, and his fellowmen for an unfulfilled promise from his employer, trust or corporation.*
>
> *Esau was a traitor to himself; Judas Iscariot was a traitor to his God; Benedict Arnold was a traitor to his country. A strikebreaker is a traitor to his God, his country, his wife, his family and his class.*

It was printed fancy, with a border around it, suitable for framing, like a dime-store Lord's Prayer. Louie picked it up from the table, read it over again, and smiled. "That man sure knowed his scabs, didn't he?" he said.

Along Clover Fork a scab may be a storekeeper, a neighbor, a fellow coal miner, a member of your church, a cousin, even a brother. But he is a member of a distinct and separate segment of the community. As other communities are peopled with different ethnic, racial, and religious groups, Clover Fork is inhabited by scabs and union men. It is impossible for an outsider to tell which group is dominant, or to recognize a man or a family as members of one group or the other. The lines were drawn years ago—even generations ago—during other well-remembered times of conflict.

As he drove along Kentucky 38, Louie was asked to stop at a little roadside store for a soda pop.

"Not that store, buddy," he said.

That's a scab store. You wouldn't want to do any business with him. He's a full-fledged scab. Why, back in that strike in '64, he sued the United Mine Workers for ten thousand dollars, I believe it was. He claimed somebody throwed a little dynamite off on him there one night. And I believe he won that case, too, best I remember. He's one of the gun thugs that goes up there of a morning to Highsplint. He went through our picket line in '64. And that right there was where he started being a louse. But from '64 on back, he had worked under the United Mine Workers for years and years and years. And he's got the time under the United Mine Workers to draw a pension. That's what makes it bad. A man like that would draw a pension from the union, and be fighting us as bad as he's fit us for the last ten years.

Naw, buddy, you don't want to do business with him. Let's drive on down to Deaton's.

Some scabs, like the storekeeper, fall, like Lucifer, from glory. Others are born into scabism. Scabness is in their genes. Along Clover Fork, these are called thoroughbred scabs.

Junior Deaton said:

If a deaf-and-dumb person marries a deaf-and-dumb person, nine times out of ten they have a deaf-and-dumb baby. I think if a scab marries a scab, they'll have a thoroughbred scab. If a scab marries a union person, they'll have a sixty-percent scab. I think it's born. I think you're born with it or without it. I really do. There has always been scab families and union families. And I think where you get your radicals is where a good union man marries a good union woman. You know, where both of them is sort of radical for the union. They'll have a kid that's really radical. I don't believe in being radical. I believe in being very level-headed and listening to what the company has got to say, or what the other man has got to say. I believe in levelness. I don't believe in going up and saying, "You gonna do this; you gotta do that." Ask them in a nice way. If they don't do it, then tell them. That's what I mean by levelness.

But being born to scab is something that you can't help. I just don't believe that you can make a real union man out of a scab. Now, some is converted somewhat. We converted some on this checkoff. But we had a hard time converting them, and now that we got them, you sure ain't gonna get them out on no picket line. They're ashamed to be caught at a picket line. I had a born scab in my outfit there during the '64 strike. I went to his house after dark one night and told him if he wasn't there at four-thirty in the morning, waiting right there at the bridge on me, that he'd no longer benefit from the United Mine Workers, which would cut his voucher off and everything. And that son of a bitch stayed up all night. He could not go to bed. He stayed up all night. But he come out on the picket line.

Old man Whitfield, him that owned the mines then, he come by that day, and that feller fell off the log he was setting on, trying to get behind it. He didn't want the man that owned the mines to see him on the picket line.

Naw, buddy, there ain't no converting no scabs. Not really. Now these scabs that worked at Brookside, and didn't *vote for the union? They'll come crawling back after the strike. They'll want the benefits. But you won't catch them at no union meeting.*

You know, I don't believe in ever saying anything around a man's kids. A scab we got lives in Ages here. He's got two kids that come to the store regular, and I don't like for nothing to be said while they're in here. But I heared one of them talking the other day. They was talking about this strike. And that little feller said, "Well, I ain't for neither one of them." And I asked him to repeat that. I said, "What did you say, buddy?" He said, "I'm not for neither side. I'm not for the scabs, and I'm not for the union men." He's calling his daddy a scab! Now he's gonna be raised knowing that his daddy's a scab!

When the men in a mine vote to join the United Mine Workers and go out on strike, the born scabs who work in the mine have two alternatives: join the strike, or go scab elsewhere until the strike is over. Even if the union wins the strike, the scab is entitled to go back to work there if he wants to, by becoming a nominal member of the union. If he chooses to go back, though, he can expect trouble.

The hurts that people put on each other during a strike don't heal too easy. These Brookside men have put up with a lot during this strike, and they don't look too kindly on scabs. Now the scabs that worked at Brookside, and then went to work at Highsplint after the strike started, I think the company ought to keep them up there. They went as scabs up there to work, and the men up there got used to them. Now put those men back down here at Brookside as soon as you get a UMWA contract, and the company would be asking for it. They're really hated. If the company wants to move them back down here, they oughta wait about a year, till things cool off. By that time, maybe the union men will have met them on the road somewhere and discussed it, talked it over with them.

Yeh, buddy, a scab'll do anything. Sometimes he can walk right by you and never see you. Never know you was there. Other times, he'll walk by and just grin and grin at you. You can spot them a mile away.

Besides the fallen-from-grace scab and the thoroughbred scab, there are at least three other subclasses within scabism: the expedient scab, the imported scab, and the gun thug.

"They put a guy in jail down in Harlan," Louie said.

Stayed in jail about all the time. Drunk and disorderly and every other kind of crime you can think of. Well, he let people know that he'd scab, if they would let him out of jail. So the company went down and talked to him and said, "We'll get you out of here, if you scab for us." Well, when they let him out, he come straight up to our picket line and told us what was going on. He said, "Now, they let me out of jail," he said, "and I'm gonna act like I'm going to work for them. Now, I'm going down to that company store, and I'm going to buy everything I can on credit, and then I'm leaving." So he goes in the store and buys a whole new outfit. You know, belts and shoes and a hard-shell hat and all the other stuff he could get. And then he leaves. Goes plumb out of the county. That tickled me to death.

That man was a *phony* expedient scab. If he had gone to work, he would have been a real one. Then there is the imported scab. "They was going plumb down into Tennessee and Virginia and way over into other parts of Kentucky to get them," Louie said.

They was getting wine-heads and needle-poppers and grass-smokers. Actually, they didn't want them for miners. They just wanted them in here to cross the picket line and break the strike. They tried to make it look like they had a big force of men, but hell, everybody knowed they wasn't no good. They wasn't about to mine no coal. The company didn't want them for miners. You gotta have a man that's willing to work before you can mine coal. Everybody knows that. The company knows that.

If the imported scab succeeds in breaking the strike, he leaves, perhaps to find another strike, and turns the mine over to homegrown scabs. If he does not succeed in breaking the strike, he leaves, too. If he has a home anywhere, it is not along Clover Fork.

Like the thoroughbred scab, the gun thug usually is homegrown. He has lived in Harlan County, and perhaps on Clover Fork, for most of his life. Perhaps his father and grandfather lived there, too. If they were gun thugs before him, then he is a thoroughbred gun thug.

Some gun thugs are company men gone wrong. That is, they are regular, salaried employees of the company and hold jobs that the miners consider legitimate. A company man may be an assistant superintendent, or a foreman, or a security guard. Strikers do not object to his crossing the picket line. But if he loads his car with scabs and tries to run them past the picket line, he will be carrying a gun. If he uses it to threaten or injure a striker, he is a gun thug.

Some thoroughbred and fallen-from-grace scabs moonlight as gun thugs. "He's a hard-working man," one striker said of his neighbor. "He scabs in the daytime and thugs at night." Gun thugs snipe, pistol-whip, dynamite, fire into houses, and sometimes kill.

In the past, at least, some gun thugs were officers of the law. During the severe labor troubles of the 1930s, the sheriffs of Harlan and other coal counties routinely deputized gun thugs hired by operators to put down the organizers and the strikers. John Henry Blair and Theodore R. Middleton, who held the Harlan sheriff's office during most of that bloody decade, deputized hundreds of the most notorious toughs in the mountains, granting them the powers of arrest and search and seizure, which were used as a general license to terrorize. The salaries of nearly all of those men—including the three slain by miners in the Battle of Evarts in 1931—were paid by coal companies.

It was such men, Louie said, who gave Harlan County its infamous reputation, which he believes has been perpetuated unjustly by the press and television.

"A lot of people who live other places, they think the people who live in Harlan County are just like Mammy and Pappy Yokum," he said.

We've had reporters come in here from New York and places like that, and they went back up there and wrote stuff like that, you know, like we done things entirely different from the way the rest of the world done them. They got those ideas back in the thirties, when they had a hard time organizing, when they used to call this Bloody Harlan. Well, I ain't denying there's a lot of people back in here that'll lay the hammer back on you if you cross them. But most of them are just like me. They'll be nice to you if you treat them right.

He paused and spat a brown stream of Bruton's Snuff into the small sand-filled bucket beside him.

Course, it's starting to look like we may go back to those ways. It sure looks like we're coming to it. If we can't get justice in the courts, if the county law and all his deputies, and the state polices and Governor Ford ain't gonna see that we get justice, they's going to drive us into the mountains. I mean, we're not going to be treated like that no more. Not in this day and time. If they're not going to let us man the picket line up there peaceful, why, we're going to have to try other means to do it. You know, go back to the Mammy Yokum days.

The Brookside miners seemed to hold no grudge against High Sheriff Billy G. Williams, a local well-driller who happened to win public office in a difficult time. Williams had dutifully served his warrants and eviction notices, but he had managed to do it without raising hackles. He also had not filled Clover Fork with deputies. "He's stayed out of this pretty well," Louie said. "He used to be a coal miner hisself. He could give us hell if he wanted to. He has the authority to deputize as many men as he wants when there's trouble. The high sheriff we had in 1964 must have had three hundred deputies. But Billy G.'s stayed out of it pretty well."

The real trouble this time, the miners said, was from the thoroughbred gun thugs, whose anti-union meanness had been passed down to them by a succession of inherently evil forefathers.

There are times, though, when even the line between the scab and the union man becomes fuzzy. No true union man would go back to work at Brookside after the 1964–1965 strike, because to do so would make him a scab. He might have to accept a job in another non-union mine, though. Such a man was a dispossessed Israelite, weeping beside the waters of Babylon, awaiting his day of deliverance.

When the picket line went up at Highsplint, the scab-versus-union situation became ticklish. The Highsplint miners had *not* voted to join the United Mine Workers. They were *not* on strike. And when their Brookside friends and relatives picketed the mine, the Highsplint men were faced with a difficult choice: either cross the line and be regarded as scabs, or honor it and risk losing their jobs. Those who refused at least to try to cross it *did* lose their jobs.

Betty Eldridge, wife of another Brookside miner and one of the more active women in the picket-line fights against the scabs, was one of many who felt uneasy about the predicament. "To be honest with you, I would have to say that I was against picketing Highsplint. I know the men has to work to support their family, and I was afraid that we might get them out and they'd lose their jobs. That's hard on men. You can't live without money coming in. But I knew that we needed to cut down on Duke's coal supply, too. Still, we know most of the people who work there, and some of them are good friends of mine. In fact, I had a cousin or two working up there. It's a hard problem."

6 THE STORY THUS FAR

In 1970, Duke Power Company went into the coal business. It established a new subsidiary, Eastover Mining Company, through which it bought the Brookside mine from Harlan Collieries. From other companies Eastover bought the Highsplint mine and the Arjay mine on Straight Creek in neighboring Bell County, and the Jawbone mine in Virginia. The coal from these mines does not go on the market, but straight into the furnaces of Duke, the nation's third largest user of steam coal.

Five days after Duke purchased Brookside, Eastover Mining signed a contract with the Southern Labor Union, which had been organized some twenty years before—with the encouragement of the coal operators, some say—as an alternative to the United Mine Workers. The Brookside miners, who had had no union at all since the failure of the strike of 1964–1965, did not vote on the SLU contract. Eastover said to them, in effect: "We are your new company, and here is your union, and here is your contract."

In May 1973, the Brookside miners began signing cards calling for a representative election before the expiration of the SLU contract in July. Several UMWA supporters were fired, but on June 26 the men voted 113 to 55 to switch unions. Negotiations between the UMWA and Eastover got nowhere, and on July 26 the Brookside men struck. This time they had the support of the international union. The Miller administration decided to pay a strike benefit of $100 a week to strikers with families—said to be the highest strike pay in the history of American labor—and to provide the strikers with medical insurance. Miller and the other international officers, Mike Trbovich, the vice president, and Harry Patrick, the secretary-treasurer, visited the picket line, pledged their support to the end, and warned against violence.

In September the strike began to take a familiar Harlan County course. Eastover started sending strikebreakers across the picket line and petitioned the Harlan Circuit Court to limit the number of pickets at the mine. Harlan Circuit Judge Edward Hill pleaded ill health and turned the case over to Letcher County Circuit Judge Byrd Hogg, who had been a coal operator himself. Judge Hogg ordered that only three strikers be permitted to stand at each entrance to the mine. Soon, up to sixty men a day were walking past the pickets into the mine.

In late September, some fifty Brookside wives drove into Harlan and staged a demonstration in front of the office of the Harlan County Coal Operators Association, the clearinghouse for employment in Harlan County mines, and, some say, the keeper of a blacklist of "undesirable" miners, such as UMWA supporters. While the women were marching, someone called out, "Let's go to Brookside!" and the marchers piled back into their cars and arrived at the picket line just as the strikebreakers were coming out of the mine.

"Our men couldn't do nothing," Ruby Stacy said. "That judge had limited them to three pickets, and they had to just sit there and watch them scabs go in and take their jobs away from them. They was about sixty or sixty-five working the day we come up there. We decided, well, if our *men* can't stop them, *we* will. We wasn't under no court order. At least, we didn't *think* we was. We picked up some switches along the river bank and alongside the road there."

Louie laughed. "Did you ever see a switch as big around as your arm?"

"We caught them just as they was coming off the hill," Ruby said. "We taken after them with the switches. They stopped dead in their tracks. They couldn't get out. They was going back up the mountain and around the mountain and every whichaway, trying to get away. They looked like a bunch of rats. It was really funny after it was all over, but it wasn't too funny right at the time."

In October, Judge Hogg ruled that his court order had applied to the women, too, and found seven of them—and nine men—in contempt of court. He ordered them jailed immediately.

Several of the women had brought their children with them to the courtroom. The children went to jail, too.

"We didn't have no idea we was going to jail," said Nannie Rainey, mother of seven, swaying slowly back and forth on the porch swing at her four-room company house.

We took the kids to court with us because we didn't have nobody to keep them. I had three of mine with me. I think there was about twelve kids in all. They put us all in a cell together. What they called the bullpen. That evening, a welfare worker come and told us that if we couldn't get rid of the kids, they'd come and take them away from us. We wasn't about to let nobody take our kids away from us, and Freda Armes, she took off her shoe and started after the guy with it. He left and said he was going to send somebody else up there, but he never did.

The kids, they was in there two or three hours before kinfolks could come and get them. They didn't say nothing. They just sit there, and the people at the jail asked them did they want anything to eat. What they had to eat was just cornbread, and nobody couldn't eat it. I noticed the guy taking it around dropped a piece on the floor, and he just picked it up and put it back on the plate. That floor was nasty, too.

Photographs of the women and children staring from behind jail bars were published across the nation and attracted public attention and sympathy to the strikers. The women received letters of encouragement from fellow housewives in many states. Some included checks. The women organized themselves into the Brookside Women's Club and began raising funds to buy Christmas toys for the strikers' children and medicine for the ill among them.

The international union stepped up its publicity campaign against Duke Power, buying ads in North Carolina papers to plead its case against the company. It also joined Carolina labor and consumer organizations in combating a 17 percent rate increase requested by Duke, and transported delegations of Brookside strikers to picket the utility company's headquarters in Charlotte. The union claimed that a number of safety violations which had been cited by federal inspectors in the Brookside mine had been allowed to go uncorrected for a year or longer. In November, federal inspectors found seventy-two violations there.

Negotiations were resumed. Eastover announced that it would not sign a contract unless it contained a no-strike clause requiring the union to pay damages for unauthorized strikes. The company also insisted on full management control of promotions, rejected a portal-to-portal pay proposal, and refused to pay the seventy-five-cent-per-ton royalty to the union's welfare fund. The company wanted to gut the union's national contract. Negotiations broke down again.

On February 28, 1974, the company suddenly tried to send strikebreakers into the mine again, and the women reappeared with their switches. Among them was Minnie Lunsford, daughter, sister, and mother of coal miners and wife of a retired miner. She said:

> *Them scabs was driving right by the pickets and spitting on them and cursing them as they went by. They had pistols on the seat beside them. Well, the miners' wives thought that was kind of pitiful, and they wasn't gonna let nobody treat their men thataway. That's how come we went out on the picket line ourselves. I was scared until it got a little bit rough, and then I got mad and wasn't scared anymore. I whupped the tar out of one of them scabs with my walking stick. A little cane, you know. It had a hook in the end. Well, I didn't walk with it, but, you know, it was made for that purpose.*
>
> *This scab come up in a car, and he had his window down. Poor feller, he didn't have no more sense than I've got. He tried to turn into that little narrow lane there by the company store, and I poked my stick through the window and got him right in the innards.*
>
> *He sat there looking at me. I guess he thought I'd move, but I didn't. I got right in front of his car. He said, "Will you move, lady?" And I said, "No, I didn't come here to move. I'm gonna stand here all day, if that's what it takes to stop you." Just one other girl was where I was at. The rest of*

them was down by the bridge. They had somebody treed down there, whupping the tar out of them. Anyway, this scab said, "I could push you down." And I said, "No, I don't think you will. You could, but you won't. You won't do that." A man just don't hit a woman in Harlan County, you see. He puckered up his mouth and made a big, ugly face at me and said, "Will you go to hell?" I said, "No, that's where you'll go, if you keep on with this scabbing." And I commenced to switching. I couldn't get a big swing at him, you know. Just in the window. I made his old neck and face red, though, and, boy, he got out of there.

Somebody asked him that evening was he going back the next morning, and he said, "No, there's some old lady down there just beat the heck out of me."

I enjoyed it. It was important, too. If it wasn't for the women, it would be Bloody Harlan now, let me tell you. Just two weeks ago, I was up at Highsplint picketing. I don't care. If they put me in jail, I won't have to stay long. I wouldn't mind going to jail. I'm no better than the rest of them, just because I'm old.

There were some brief scuffles with the State Police, new contempt-of-court charges, and more women went to jail. But no further attempt was made to work the Brookside mine.

About two weeks later, an independent Citizens Public Inquiry into the Brookside Strike, which included on its panel former Secretary of Labor Willard Wirtz, former United States Senator Fred Harris, and seven other liberal educators, clergymen, and the like, set up shop at the Evarts Multipurpose Center and heard two days of testimony about safety conditions in the mine, the miners' court hassles, and living conditions in the Brookside camp.

Junior Deaton's younger brother Darrell, who works at the Brookside tipple, was among the first witnesses. He testified:

Throughout our lives and the lives of our fathers and grandfathers before us, companies like Duke Power have come into these mountains looking for wealth and riches. They took our coal, and they took our timber. They took our health, and sometimes our lives. But they never gave anything in return.

Isolated from the rest of the country, we in these mountains fought back as best we can. But we faced multi-million-dollar corporations that held a stranglehold over our lives and our land. Duke Power came to these hills only four years ago, but Duke Power isn't new to Harlan County. We've seen its kind all our lives.

All our lives, we've seen companies like Duke, which run mines for profits, but don't care about our lives and how many of us go home battered and broken. All our lives, we've known companies like Duke Power, which buy up a coal camp community like Brookside and think they've bought the lives of the people who live there as well. We know all too well about companies like Duke Power, that claim to be a good neighbor in the regions where they are known, but who come into our land under a different name to plunder and steal without a second thought, except to avoid getting caught.

A spate of articles by members of the panel appeared in national magazines. A month after the inquiry, a group of Brookside miners, including Darrell Deaton, picketed the New York Stock Exchange on Wall Street, urging investors to "dump Duke." The miners were taken to a cocktail party, where Darrell saw real silverware for the first time.

Later in April, NLRB Administrative Law Judge Maurice Bush blamed Duke and Eastover for the breakdown in negotiations. The company's insistence on a no-strike clause, he said, was a deliberate attempt to avoid coming to terms with the UMWA. In early May, two officials of the Southern Labor Union met with strikers Carl Noe and Ron Curtis, and offered to pay them $5,000 if they could get the men back to work. They paid an installment on the spot. Earl Dotter, photographer for the *United Mine Workers Journal* who was hiding in the bushes, photographed the meeting, and it was recorded by tape recorders hidden under the clothing of the strikers. The UMWA filed charges with the NLRB, accusing the SLU and the company of conspiring to break the strike. Nothing came of it, though, for apparently it is not against the law to bribe a coal miner.

In June, Eastover fired many of the strikers and gave eight families thirty days' notice to vacate the company houses in the camp. In early July, the picket line was thrown up at Highsplint, and Minard Turner was shot. The people along Clover Fork began recalling the Depression days, when John L. Lewis's attempts to organize the miners of Harlan County led to dynamitings, gunplay, kidnappings, beatings, and starvation in the coal camps. Brookside miners claimed to have been shot at by snipers along Kentucky 38, and could point to bullet holes in their pickup trucks. Men who formerly had been friends, like Lawrence Jones and Billy Bruner, became enemies. Neighbors refused to speak to one another. Children exchanged catcalls and jeers in the school yards of Ages and Evarts. Brothers tried in vain to talk one another into changing sides, and became bitter when they failed. Retired miners, living on UMWA pensions and federal and state black-lung benefits, tried to pass on the union faith to their offspring. Sometimes they succeeded. Sometimes they did not. Volleys were fired in the night from the highway down toward the company houses in the river bottom. Small children became accustomed to dropping to the floor and lying still, but did not get over their fright at the noise and the expressions on their parents' faces. One hot night in August, Mickey Messer's home, near the center of Evarts, was under fire for ten minutes.

"I was asleep when it happened," said his wife Lora. "I'd just dozed off when the shooting started. I rolled out on the floor. The baby sleeps with me, and the other three kids sleep in another room. One of the bullets hit next to the window of their room. Mick came in and said, 'They're shooting into the house. Keep the kids down.' So I got the baby, and got the other kids out of their room, and took them to the back of the house. We laid down until the State Police came and took us out of the house, up to my mother's.

"Maybe it was just because I was asleep, but it sounded like a war."

The house, an outbuilding, and a large elm in the yard were pocked with bullet holes. State Police estimated that 150 shots were fired. They found a .45 automatic on a hillside not far from the house. It belonged to a neighbor of the Messers, who said it had been stolen from him.

"We stayed at my mother's every night for a couple of weeks," Lora said.

> *I'd come back here during the day, and then take them back up there at night. There had been a lot of threats, since Mick's so active in the union. We didn't want to take a chance on somebody coming back to finish the job. All the time I was at my mother's, the kids, they would get sick at night and have nightmares and wake up hollering. And since we come back down here, they won't sleep by themselves. They still talk about it. They realize what danger they's in.*
>
> *My dad, he put thirty-three years in the mines. He was a strong union man, all the way. My mother said he'd been right there by Mick's side through all this if he'd been alive. He died, it'll be three years ago this November. He had black lung. He was disabled.*

On July 21, Arnold Miller and the other international officers had come to Harlan and addressed a crowd of several thousand miners and their friends at Cawood High School. Again, he had warned them to avoid violence. A congressional committee sent staff members to Evarts to look at the bullet holes in Mickey Messer's house. Then Miller called for a nationwide shutdown of union coal mines August 19–23 as a memorial to victims of mine disasters, black lung, and coal company violence. On August 22, several thousand miners from states as far away as Kansas were bussed into Harlan to parade through the streets. They carried a huge banner bearing a slogan of Mother Jones, an early organizer:

PRAY FOR THE DEAD
FIGHT LIKE HELL FOR THE LIVING

7 TUB & NANNIE

He is not fat, but Jerry Rainey has been called Tub since he was a child. The nickname is so old that he does not remember how he got it. He has the clean-cut, innocent look of a small-town high school football coach. He is thirty-five years old. So is his wife Nannie. They are children of coal miners, and grew up together in the Brookside coal camp. They went to school together at Ages and got married when they were twenty years old. Now they are parents of six daughters and a son, ranging in age from five to fifteen years. In August 1974 their seventh child started to school.

The Rainey family rents a four-room house in the Clover Fork bottom from Eastover Mining Company. They pay $10 a month, and $10 more for utilities. There is no plumbing in their house, but the Rainey children are bathed every night and their school clothes would do honor to one of the detergent commercials that flicker, day after day, across the big color TV screen in the living room. The tiny house is spotless. Nannie spends hours every day carrying buckets of water from a faucet about thirty yards from her front door and heating it on the stove in her small kitchen. She has a look of perpetual weariness about her, but she smiles a lot. So do Tub and the children. Their individual cheerfulness, their closeness as a family, and the beauty of the children attracted many of the newspaper reporters and television crews that tramped through the camp during the strike. They were interviewed and photographed many times. By late August, both Tub and Nannie and three of their children had spent time in jail, and the Raineys had received their second notice that they were about to be evicted from the camp. Nannie said:

This ain't a particularly good house we got. It ain't much of a house atall. We wouldn't mind moving, if there was a place for us to go. But there ain't no place to go in Harlan County. There ain't no houses that we can afford to rent. Even if we had the money, nobody will rent a house to you if you got kids. Dogs and cats is okay, but they don't want kids.

The company says they're making everybody move so they can tear down the houses, because they ain't no good. Well, that's true. They ain't no good. But they're all we got. And the company ain't evicting nobody that ain't active in the strike. They've just come in here with men and moved some people's furniture and stuff out on the ground, and then start right in tearing down the house. I kind of felt sorry for old Billy G. when he brought this second notice. He stood out here on the porch and said, "I'm sorry, Mrs. Rainey. I hate to bring this to you worse than anything." I said, "Lord, that ain't nothing. I can take and use it for toilet paper."

She laughed, either at her joke or at Billy G.'s embarrassment, or at life in general.

The other women, they say that if the company comes and tries to set us out, they'll come and carry the stuff back in the back door while the company men is carrying it out the front. When I move, I'm gonna paper one of them rooms with UMWA bumper stickers, buddy. I'm gonna have me a UMWA room before I leave. Because that's all these notices are. Just another way to make things hard for the union men. They know Tub and me, because we was born and raised here, and every time they want to cause trouble, they say, "Go get Tub and Nannie."

Tub, unlike many family men, decided not to seek a part-time job during the strike, but to get by on the $100 a week the union provided him. "Every day there's been a picket line, I've been there, either at Brookside or at Highsplint," he said. "I just wouldn't feel right, drawing a check from the United Mine Workers, and then going off and working somewhere else while we're on strike. I'd rather be on the picket line, even though I don't *have* to set but one day a week. I like to be there where I can see what's happening. And if anything does happen, I want to be there when it happens."

One bright morning Tub was sitting in his old red pickup on the Brookside picket line, whittling a cedar log and talking to the other pickets, when he looked up and saw Nannie and their seven children walking down the L&N tracks to school. He got out of his pickup and stood watching them, grinning. The sun glistened in their blond hair. They grinned back at him. The little boy, wearing a new necktie, gave him the widest grin of all. "You're looking good, buddy," Tub told him.

Nannie had managed to save enough out of the strike benefit checks to buy each of the children two new outfits for school. "I don't want them to be laughed at," she said. "There ain't no reason they can't be as proud as the next kid."

"Yeh, I'm proud of my kids," Tub said.

I mean, a man can't raise a family without them, and I'm just proud that they're all healthy, and they ain't none of them lost no weight during this strike. They're all healthy and active, and that's enough to make a father proud. One of them, the next to the oldest, took the mumps last winter, and as soon as she was over it, the other six got it. That was pretty tough, having six with the mumps at the same time, because the doctor said to keep them quiet and don't let them be running and hollering. We had a hard time with that, here in this small house. They wanted to run and jump and holler and play ball. But they pulled through it, and they've all been healthy.

We all get along good with each other. And they know what's going on. They've heared the shooting. They've had to get out of their beds and lay down on the floor. Even my baby, when she sees a scab drive by on the highway up there, she'll say, "Daddy, there goes a scab." She knows them, see? She knows what they are. She watches for the bumper stickers, you see. She'll see a car pass with one of them union stickers on it, and she'll say, "Daddy, there goes a union man." She sure does.

I want good things for my kids. I've got about seven years in different mines, and I've never been a member of the United Mine Workers until now. But I remember when this place was UMWA. I was born here, and my dad was working under it. My brothers worked under it. All my kinfolks was working under it, and I know what it can be. I've worked where they didn't have no union at all and was treated better than what we was treated under the Southern Labor Union. I used to work in a plating plant up there in Michigan. They didn't even have a union, and the company up there was better to us than what they was here under the Southern Labor Union. Every year, we got a big Christmas bonus. They throwed a big party every Fourth of July and invited all the families. We couldn't have been treated no better by a company.

My dad, he's still working up in Michigan. I can remember when I was just a little kid, he'd go into the mines right here at Brookside in the morning before daylight and come out after dark. I can remember him carrying his dinner bucket. But things was better here then than they are now. My father-in-law, he's getting a UMWA pension. They've always been union men. Nobody gets a pension from the Southern Labor Union, buddy.

Tub's seven-year-old boy says he has already decided to follow his father's footsteps into the mines, when the time comes.

Course, it's a long time off, before he'll be ready to go to work. But the way I feel now, I wouldn't want him to go into the mines, the way things are, with all this scabbing, and this Harlan County Coal Operators Association that blacklists you, and this Southern Labor Union with nothing but company sucks for officers. But if they get Eastern Kentucky organized under the United Mine Workers again, and get them a good safety program set up, which they ain't got; if

> *the mine is run safe, like it should be run, I wouldn't worry a bit about him going in the mines. I* like *working in the mines. And if they run the mines safe, they wouldn't be much more danger working in the mines than getting in an airplane and flying every day.*

Nannie, sitting on her porch swing, gazed out toward the clump of weeds hiding the faucet from which she bears her water. It was almost time to start on the laundry. "I didn't think the strike would last this long," she said softly, as if to herself. "But I'd rather for him to stay out than to go in there without safety. Because if he goes in there and gets killed, then I won't have nothing, you know. There ain't *nobody* to help me and these children. With seven children, you've got to have somebody working. But I'd rather for him to stay out than to go in there and be a scab. Anybody that goes down that deep, you don't know whether they're going to get back out or not."

8 THE MAINEST THING

Coal mining is the most dangerous work in America. It kills many men, sometimes by the dozens in explosions and cave-ins, regularly by twos and threes in machinery accidents and small rock-falls. Hundreds more each year are "maimed up," as the Brookside men say. And even if a miner stays alive and whole until retirement day, he is almost certain to have "black lung," one or another of the respiratory ailments caused by the coal dust and rock dust that settle in his lungs during his years underground. Miners, when they count up the money they will have for their retirement years, routinely figure in disability payments from state and federal governments for black lung eventually, when breathing just becomes too much work.

Yet, few of the Brookside men want to leave the mines to go into another line of work. "I've always wondered why people likes the coal mines, once they get in them," Junior Deaton said. "It's not the money. They could make as much or more doing something else. But they's something about coal mining. I believe it's just that the old coal miner, he thinks he can't do nothing else. He can't make it nowhere else."

Perhaps, in the beginning, many of them felt like Jeffrey Day, who, at nineteen, has been in the mines for two years. "The first time I went into a mines, it was pretty spooky," he said. "You know, they was drilling and shooting, and the first time they shot off that powder, it like to scared me to death. I thought the whole damn thing was falling in, the whole top was coming in. It took me about a week before I knowed whether to run or set still. After about the first week, though, I was settling down pretty good. And then, when I come up here to Brookside and found out how bad that top was, why, I was skittish for two or three months. I always thought that mountain was about to fall in."

After a while, though, they forget their fear, as Louie Stacy has in the thirty-six years since he first put on a miner's lamp. They accept the danger as a normal aspect of life. "A coal mine ain't no worse than anything else," he said.

> *You just have to get used to working in the dark. You've got this light on your head, and you can see a place about thirty-six inches wide straight ahead. Both sides of you is dark. And in the coal mines, it's darker than it is out here when it gets night. That's the darkest dark there is. You can't see your hand two inches in front of your nose. But you never feel like you're alone down there, even in the dark. You can always look around you and see the lights of the other men.*
>
> *I've left coal mining a few times when times was bad. I've learnt other jobs. I've learnt carpenter work some, and I've worked some in a steel mill. But it just wasn't the same as coal mining. I just wanted to come back to coal mining. Cause I've worked so many years at it, I guess. See, I worked in the mines for years and years and years before I ever got out of it. And it just gets in your blood, I guess. You just want to come back to it.*

Mining has changed a lot since the days of the ponies and the carbide lights and the necessity of "growing muscles on top of muscles" in order to survive. Today, in the larger mines, even the blasting powder is gone. It has been replaced by the continuous miner, a huge machine resembling a praying mantis. It moves ponderously up to the coal face on steel tracks, like a tank, and gnaws the coal out of the mountain in small chunks. The coal moves out of the continuous miner on a conveyor belt and is dumped automatically into shuttle-cars, which carry it to the preparation plant outside at the tipple. After the continuous miner has gnawed several feet of the mountain away, it backs out and moves into an adjacent working place to do its job again there. The roof-bolt machine then moves into the place the miner has abandoned, drills deep holes into the rock overhead, and inserts long bolts to pin the mountain back together.

In this machinery, the energy industry comes full circle, for it is powered by electricity. Kentucky electricity provides the power to produce the coal that will produce North Carolina electricity.

The machines require constant vigilance and maintenance if they are not to turn against the men who use them. Proper maintenance requires money and time, and, for the short run, at least, slows down coal production, which slows down electricity production. The reluctance of many coal operators to keep their mines and machinery as safe as possible has led to the passage of rather stringent state and federal laws requiring them to do so. The laws are enforced by inspectors, who pay periodic visits to every mine, inspect it, and cite the operators for any violations they find. The United Mine Workers national contract of 1971, which Eastover would sign if the Brookside men won their strike, also provided for a safety committee whose members were to be elected by the miners themselves. Under the contract, the committee would have the authority to shut down a piece of defective equipment until it was repaired, or even to order the evacuation of a mine when large-scale danger was imminent.

For the UMWA, the safety committee was an unnegotiable demand. The issue was self-determination at its most basic level—the right to stay alive without the consent of the boss. The Southern Labor Union contract at Brookside provided for no safety committee, and Norman Yarborough, president of Eastover Mining, insisted that he would never sign a contract that would give the miners the right to shut down his mine. Such a decision, he said, is the prerogative of management alone.

Tub Rainey operates a continuous miner, a top-pay job bringing him forty-two dollars a day under the SLU contract. It is also one of the most dangerous.

"I've been in the mines a lot when the inspectors come in," he said.

The bosses knows ahead of time when they're coming, and they tell you when we start in of a morning to watch yourself and get everything straightened up, that the mine inspectors is coming. They'll tell you the names of the ones that's coming.

The miner runs on a big electrical cable, you know, about as big around as a man's arm. It carries four hundred and forty volts of electricity. I've been in there three or four times when that big cable would blow, and the inspector standing right there. You know, to do a good splice job on that cable takes a good forty-five minutes to an hour. And the boss would say, "Don't bother to splice the cable. The next shift will take care of it." And the inspector standing right there. And the next shift wouldn't take care of it.

That miner has a big power box on it, and the man who's operating the miner sits right next to it. You're right up against it when you're running the miner. The electricity runs through a rectifier in there that has big circuit breakers in it, and it's supposed to automatically shut off the juice if something happens. Well, the rectifier on this miner had been hot-wired, and a buddy of mine was operating it one day, and that big power box just blowed up. It blowed a big hole about three inches in diameter right in the back of that power box, with this man sitting right up against it. They was two federal inspectors and a state inspector in the mine the day that happened. Well, the bosses said they didn't have another power box, that they'd have to order one. And they hot-wired the power box again. That happened eight months before we come out on strike, and when we come out, that same box was still on the miner. If we'd had a United Mine Workers safety committee, that miner would never have been started again until a new box had been put on there.

One day about a week later, a buddy of mine, Jimmy Osborne, was running that miner, and I was helping him. That cable—it's about three inches in diameter—it blowed in two. When it does that, that rectifier breaker's supposed to cut the power off. Well, it didn't, and that big old cable, it stood up *like a big black snake, and fire was shooting everywhere. It run me and Jimmy up to the face of the coal. We couldn't get around the miner to cut off the juice. We started squealing, and there was a boy farther down from us, and he heard us hollering and he run and shut the juice off by hand.*

Tub paused and rubbed his hand over the bristles of his crew cut.

I guess safety is the mainest thing this strike is about. I can't help feeling that a lot of these inspectors is just playing along with the company. We can't depend on them to keep things safe for us. Like these dust samples. I run a miner, and that's a dusty job. I get a lot of coal dust. And they have these little old machines they're supposed to use to take dust samples in the mine, for the inspectors. The boss'll give you one of those little machines and say, "We've got to get a good sample today. If we don't, we'll have to carry these damn things a week, you know." They give me a machine, and I go to cutting and loading with the miner, and the boss'll come up and turn the machine off. He done that for a while, and then I got to thinking about it, and every time he would turn the machine off, I'd turn it back on. And he'd yell, "You're gonna get a bad sample and get the damn mine shut down!" And I told him, "Get the damn air up here like it's supposed to be, and get the water on the miner, and there wouldn't be no dust." The miner's supposed to spray the coal face with water to keep the dust down, but it always just came out in a little trickle. That's what them dust machines is for, to protect me. I turn that off, and in a year or two something happens to me, and I get down real bad on rock dust, and they go through them samples, and I got good samples all the time. I wouldn't get no compensation.

They'd put those machines on some men in the morning and run them until they got inside the mine, then they'd take them off the men and put them where there wasn't no dust and let them run until the batteries run down. Then at quit-

ting time they'd put them back on the men, like they'd worn them all day. They didn't like me, because I'd wear mine all day. They claimed I got bad dust samples all the time. Well, I was in bad dust every day. All they was interested in was running another ton of coal, just as fast as we could. They didn't want to take the time for no safety.

The roofs in the mines operated by Eastover Mining Company vary from about thirty-six inches high to about forty-eight. The men must squat or kneel under them all day. The Brookside mine has a reputation for "bad top." Much of the rock over the men's heads is faulted and threatens to fall when the coal which supports it is cut out from under. As the continuous miner works, the roof is propped temporarily with timbers until the roof-bolt machine pins the layers of overhead rock together with bolts that vary from three feet to about eight feet long.

If Louie Stacy and his roof-bolt machine do not do their job properly, the mountain may cave in on the miners. Louie worried about his machine.

The federals says them bolts is supposed to torque at about a hundred and fifty to about a hundred and eighty pounds. You know, torquing tells you how tight the bolt is. And if they torque at more than that, why, you've stretched the bolt, and the bolt is no good. A little weight hits the top, it's liable to come in on you. Well, anyway, the bolting machine I was running was torquing them bolts at three to four hundred pounds. That was right at the breaking point.

Now, I know that I was doing wrong, but I couldn't get the company to fix my bolting machine. I mean, they'd make out like they was doing something, but it was right back to the same old stuff again. And when the federal inspector would come in—they always knowed ahead of time when the inspector was coming—my section foreman would tell me, "Now, when that inspector gets in here, don't be running your bolting machine. You jim-whack around doing some other kind of work—setting timber, or mucking track or something—until he gets gone." I had to do that. I knowed I was doing wrong, but I had to do that in order to hold the job. See, under Southern Labor, if I'd refused to do that, they'd just fired me. And that's one reason we want the United Mine Workers.

You know those shelly-cars that hauls the coal from them continuous miners? I've seed them cars so hot that them boys would have to get in hunch positions, you know, up in the seat, and jump in and out of them cars. Cause, you know, they couldn't set one foot out on the ground. If they did, they'd be almost shocked to death. There'd be a hot wire laying against the frame or something. The bosses would make them go ahead and run them. Oh, there was lots of safety violations in that mine.

Under the Southern Labor Union contract, Louie's daily pay did not begin until he had traveled five miles underground to his working place. The hour-long round trip through the darkness every day was on his own time. Sometimes, on the way in or out, rock that had fallen since the last trip blocked the track. The men would have to pile out of the car and shovel the rock off the track before they could continue toward their destination. That work, too, was on their own time.

Darrell Deaton does not encounter such problems. He has never worked underground. He is a repairman at the tipple, the preparation plant on the mountainside where the coal is washed and prepared for shipment to Duke Power's furnaces. But he has had problems of his own.

You're not supposed to work alone, ever, either in the mine or in the tipple. There's just too many things that can happen to a man who's working alone. But I repaired by myself for six months. At that time, I didn't even have telephone contact with anybody outside. I had radio contact with the office, but there wasn't nobody in the office of a night. Now, every inspection, my job was to get the tipple ready for inspection. They always knew when the inspectors was coming.

But there just simply wasn't enough hours in the day for me to get that thing safe. The major things I did, the minor things I didn't. I helped the company get away with a lot of violations. We had a lot of rebuilding and remodeling going on at the time, and we had contractors in there doing the work. If the inspector spotted a violation, like an electrical cable going across the floor or something, I'd say, "Oh, that's just temporary, while the contractors are in here."

Now, if the United Mine Workers had been in here, I wouldn't have been working alone the night I got my arm caught in that belt. I probably wouldn't have got my arm caught in that belt. Course, that was my fault. Nobody told me to stick my arm in there. I'd been called in to this special repair job, and if I'd had somebody helping me, I could have finished it in four or five hours. Anytime you work alone, it takes you a lot longer to do something.

But I had been working since about eight o'clock in the morning, and it was about two o'clock the next morning when my arm got caught. I was aggravated, I was tired. I didn't even take no lunch up there, cause I thought it was just a little job. I wasn't thinking right, and I stuck my arm into that belt, and it pulled me in right up to my chin. I was there for forty-five minutes before the night watchman found me. If I'd passed out, my weight would have flipped, and it would have pulled my arm out of the socket. But I never. I hollered about a half a minute, and then I didn't have no voice. I was scared, you know. My mouth got dry, and I just couldn't talk. Then I finally got to where I could holler again. Normally, my voice don't carry too good. But it seemed like I had a little something extra. The night watchman finally heard me. We didn't have no phone service, so he had to go wake up the boss. They called the rescue squad at the hospital. They stood and held my feet, while some men cut the machinery with a torch to get me out.

I was out for eight weeks. It wouldn't have happened if we'd had the United Mine Workers in here. I'm convinced of that.

If they complained too much about safety conditions, or pointed out violations to the inspectors, the miners said, they were moved to less desirable or more dangerous jobs elsewhere in the mine, or they were fired. They did not want to be fired, for jobs are scarce along Clover Fork, and they did not want to make that desperate journey up Interstate 75 again.

9 NORTHWARD TO EXILE

You take Kentucky 38 to Harlan, then U.S. 119 to Pineville, then U.S. 25E to Corbin, and you hit Interstate 75. Interstate 75 South goes to Knoxville, Tennessee, and then on down into Georgia. But a man looking for a job never turns that way. He heads northward, past Lexington, to Cincinnati, the gateway to the furnaces and assembly lines of Ohio and Michigan. Nearly everyone on Clover Fork has made that pilgrimage. Some stopped in Cincinnati. Some went on to Dayton. Most continued on, all the way to Detroit.

The journey is never for pleasure, but because things are not right in the mountains. A strike has failed, a man has been blacklisted. The coal market is in a slump, the operators are not hiring. A boy is determined not to follow his father into the mines.

At times the mountains empty their people onto the highway as if through a funnel. And on holiday weekends the funnel reverses itself and pours them back into the mountains again, for a few days. No one heads northward intending to stay. When times get better on Clover Fork, or when somebody at home needs them, they say they will be back. Many never come, except to visit over the Fourth of July or Memorial Day. But the coal industry is prospering now, and not so many are leaving. Nearly everyone living on Clover Fork came back from somewhere.

Not all of them took Interstate 75. Junior Deaton found work in a coal mine in West Virginia after the United Mine Workers were defeated at Brookside in 1965. Tub Rainey worked for Sprinkle Sugar in California and Revere Camera in Chicago. Jerry Johnson worked in Chicago, too, at a polyurethane plant. But sooner or later nearly all of them wound up in Ohio or Michigan, some of them several times, sometimes for a few months, sometimes for a decade.

Betty Eldridge's husband Joe is an underground electrical repairman in the Brookside mine. For seven or eight years, he held a similar job in a Detroit factory, then came home to take care of his father, who has cataracts, and Betty's mother, the widow of a miner who was shot to death when Betty was a child.

Relaxing on a shady patio in the midst of a neatly groomed lawn and flower garden behind her trailer home, Betty pondered the exile in the North, and how it changes people in ways that coal operators do not understand.

"The company is trying to deal with this strike in the same way operators have always dealt with strikes in Harlan County," she said. "They think nothing has changed, that they still own us. But things *have* changed. For years, people were stuck back here in these mountains. They never got enough money ahead to get out, you know, to see how the other part of the world was. But *we* got out of these mountains, and *we* seen that people in other places lived different. We went up there and got jobs and made money and were able to buy things that we never had here. Then, for one reason or other, we come back here, and we *still* want to live good. We won't let the operators mash us down anymore, the way they used to."

When Louie Stacy left for Detroit in 1965, he went in bitterness. His children were hungry and ill-clothed, his union had failed and then deserted him.

"You had two choices," he said. "It was either go up to Brookside and scab for the Whitfields, or just leave.

And I told Ruby I just couldn't go in there and scab. I said, "They's a big world out there, and they has to be something out there for a man." So that's when I took off for Detroit. I tell you, if you're willing to work, you can find a job somewhere. A lot of people get out and run around all over the place and claim they can't find a job. I wouldn't be a bit afraid to start off right now. Out there somewhere, there's a job you can get, if you're willing to work. But you've got to be willing to work, *you know.*

Well, they was a guy that had lived down here, and he was in the construction business up there in Detroit. His father was with us on that strike in '64, and he was telling me about this son in the construction business. We'd been out about a year then, and he said, "I'll get in touch with him and see if he needs anybody." And he said, yeah, he was needing a couple of men. Well, at that time, we had a very little bit of money. I didn't even have enough to catch a bus to Detroit. I would have had to thumb. So I told his daddy, "If he's really needing somebody to work, tell him my circumstances and ask him if he'll come down and get me." And he did. He come down here and got me. He took me up there and got me a room and advanced me money for the first week. So that's how I got started.

But Louie, accustomed to working in the never-varying climate of a coal mine, did not take to wintertime carpentry in Michigan.

We'd get up on the roof and start to put them shingles on, and that little old fine snow was just aflying. It's hard to use a hammer and nails, you know, with gloves on. I'd have to take my gloves off and rake that snow back and find out where to nail that shingle on at. I thought I'd just about freeze to death. And you had to wear so many clothes you just couldn't hardly move.

Louie quit after a year and went to work in a steel mill. For $5,000 he bought a house almost within the shadows of the skyscrapers, and spent another $6,000 fixing it up.

I never did get to like Detroit. It wasn't the people. They's a lot of hillbillies in Detroit. I guess if you took all the people from Tennessee out of it, there wouldn't be nobody left in Detroit. It got so if I wanted to talk to somebody, I'd just ask them what part of Tennessee they was from. But there was just too much of a hustle and bustle. Everybody's in too big a dadgum hurry. And you can't look out there and see these mountains like you can here.

Now, if they had Northern Steel setting right here in this bottom, why, I think I would prefer to work for Northern Steel than I would mining. But it was just where it was located at. I guess that had a lot to do with it. I just never did like a place where they's a whole lot of people. I mean, I'd prefer to be off where it's quiet. Ruby calls me old-timey, because I like it quiet, you know. I don't see anything wrong with that, do you?

Cricket jumped up on the patio glider beside Louie. Louie rubbed the tiny dog behind the ears. Cricket closed his eyes and sighed.

Ruby said, "Well, I was born and raised in these mountains, and I sure wouldn't mind leaving them."

Louie pointed toward Kentucky 38. "See that road out there? It'll lead you straight back to Detroit."

"I liked the conveniences," Ruby said.

"She liked the conveniences!" Louie scoffed. "Tell me what you had in Detroit that you ain't got right here."

"Everything. Banks, the post office, doctors. Everything. And within walking distance."

"I didn't see no doctor's office there in walking distance. You rode a dozen different busses trying to get to it."

"Why, down at the clinic! I walked there all the time!"

It was an old argument, carried on almost ritualistically, in good humor. There was a time, though, when it was not so funny.

When I heard that the Whitfields had sold Brookside to Duke Power, I decided to come back. I thought it would be a chance to start over, and maybe even get the union back. I told Ruby I was coming home, and she said, "I ain't going." And I said, "Well, suit yourself. I'm going. If you want to come join me, you can. But don't wait too long. I might get myself another woman." I come on back down here and got me a job, and got me a house at the camp.

In about six months Ruby come back, too. We sold our house in Detroit for eleven thousand dollars, about what we had in it. To a colored woman, on welfare. In Michigan, the state buys houses for people on welfare. Don't that beat all?

After five and a half years of exile, Louie was back where it was quiet, and he had the mountains to look at. Now, four years after his return, he was trying to get his union back, too.

His sojourn in the North is typical of the experiences the Clover Fork people had there. Some did not stay so long. Others made the trip several times, trying Toledo, perhaps, or Dayton or Akron or Cincinnati in various years. But few traveled as far or did as much as Darrell Deaton, before he went to work at the tipple.

He is a short, wiry man, an expert squirrel hunter. Sitting in his easy chair in his well-furnished living room with paneled walls, and children preparing for bed, and an open Bible on the coffee table beside a vase of plastic flowers, he recounted an odyssey as torturous as Ulysses' return from Troy.

I was born about a hundred yards from where I now sit. That was on January the sixteenth, 1932. My father and mother live about a hundred yards from me now. So does my brother Junior. The year I started to school, they built the school they now have here at Ages. The WPA men built it. At that time, it was an eight-grade school. Brookside was completely full, and the miners had big families. I guess they had a hundred houses back behind the store down there. They're all gone now. Now there's nothing left of Brookside, and half or more of Ages is retired people. They have only six grades at the school now, and the teachers double up, teaching two grades apiece. But back then we had eight grades and one teacher for each grade. In fact, we really had nine grades, because you started out in what they called the primer, and if you pro-

gressed real good in the primer, you skipped the first grade. About half of us skipped.

I skipped the seventh grade, too, but I wouldn't recommend that. That made it tough on me in high school. When I started to high school, I really wasn't prepared. I give it about three tries, and wasn't too interested in going no way, so I quit. I joined the Army the day I was seventeen. I was in Japan when the Korean war broke out, so I got in on the first part of that. I stayed in Korea eleven months, and then got out of the Army in June of '52.

I went to Cincinnati and worked for Kroger Manufacturing Company. Cooking candy. By the ton. I made about four and a half ton of candy a day. This was hard candies. We just cooked it all one way, and then it was flavored and colored and so on. The job I had was just cooking it. Cooked it in five-hundred-pound batches. I didn't like that job. So I come back and drawed unemployment for six months.

Then I went to Detroit and worked at Hudson's. At that time, now, when you went to Detroit, you worked there ninety days. Just before you got enough time in to draw unemployment, they'd lay off anywhere from fifty to five hundred men. Then, the next day, they'd hire that many new ones. So I worked there about ninety days, and then came back to Ages and laid around awhile. Then I went back to Detroit and worked at Briggs. That's where they make Plymouth bodies. And the same thing applied there. You worked ninety days, and that was it. It was easy to get a job at that time, though.

I finally come home and went back to high school under the G.I. Bill. I went to classes at Evarts High School, just like I did before. I think the government paid me ninety bucks a month. Or a hundred and twenty. I finished high school in '56, and then went to Dayton. At that time, jobs was getting tough to get. Well, I had a weight problem. I was short and small, and that knocked me out of two or three jobs. I think that their insurance compensation, or whatever they had, required you to weigh a hundred and forty-five pounds, and at that time I weighed about a hundred and thirty.

That's when I ended up working in restaurants. They's about a dozen of us boys from here, and some of them was working in restaurants, and that's how I ended up there. A dollar an hour. Sixty hours for sixty dollars. Now, that's just about as hard as coal mining, that restaurant work is. I tried that for a while, then come back here. I was single during this period, now. I didn't get married till I was thirty. So I come back to Ages and worked in this store my brother's got down here. That was before Junior had it. I worked there almost two years.

Then I went back to Dayton and worked in a bakery. That was another hard job, but it was pretty decent pay at that time. Two dollars and a half an hour, I think. I done it all. I started out taking bread offen the wrapper. You know, the machine that wraps it. But I was too short for that job, so I got rid of it after about two weeks. Then I got on the rolls gang, and then on the cake gang. I stayed on the cake gang until I left town. That would have been in '62. In the meantime, I got married in '62.

The reason I left, I had a skin cancer about the size of a dime, here under my eye. I didn't have enough insurance or money at that time to go in the hospital in Dayton, so I brought my wife here to Mom's and went to the V.A. hospital in Johnson City, Tennessee. And I stayed over there two months.

I got out of there in November of '62, and went over to Newport News, Virginia, where my sister lives, and put in an application at the shipyard. They had a welding school that was sponsored by the shipyard, and at that time I found out what I could do best. I was just a natural-born welder. With eight weeks of schooling, you'd normally come out and just barely be able to hold your own. But that's one thing that just come natural to me, is welding. I went through seven raises there in less than two years. If you was first-class in that shipyard, buddy, you was first-class. They didn't give it to you. You had to be able to do the job. Stayed there till '67.

Come back here and got a teaching certificate from the

state and taught one term of welding at Harlan Vocational School. That was under one of those manpower training programs. It wasn't regular high-school teaching, cause you have to have a couple of years of college for that. It was a good thing, and it paid five dollars an hour. But it really wasn't a permanent job. If you got funds, you got a class. If you didn't get funds, you might be out for two or three months. But in the meantime, I was doing some welding for a man, and when the funds played out, I just went to work for him full-time. That got me my job at Eastover.

At any time back in them years, I could have went to work in the mines. But I never did think I would work in the mines. And then, when Brookside went non-union in '65, I wasn't about to go down there then. My dad had retired as a UMWA man, and my brother had left because it went non-union.

My dad retired in '59. They had a strike in '59. The bottom was falling out of the coal market. Coal was selling for almost nothing, and the union was going downhill. The union made a deal with the company and gave them a sweetheart contract, just so they could still get the royalty for the union pension fund. They let the company treat the men any way they wanted to. And that's how they really broke the union at Brookside. That was the beginning of it, in 1959. So when contract time come around again in '64, that was it. The company just wouldn't sign. My dad refused to go back in '59. He just retired. And Junior didn't go back after the '64 strike. So, I didn't think I'd ever work at Brookside.

My dad always preached to us not to go into the mines. He never did really want us to do it. He said it's just slave labor. It's too unhealthy. So I stayed single till I was thirty, and I really didn't have any reason to go into the mines. I knew that if I learned some kind of trade that I could use in a shop or a preparation plant, I could get a job in Harlan County.

I never went away from here with the intention of staying. The longest I stayed away was four years at the shipyard in Virginia, and it surprised me that I stayed as long as I did. But we had two small kids. The last two years I worked in Virginia,

I was repairing submarines, and it was all heated work. The temperature was running around a hundred and fifty degrees minimum all the time. Course, we had blowers on us all the time, but I got fed up with it, anyway. I told my wife, "Eunice, I'm going back to Ages, put my kids in the Ages school, and stay there."

I never really had any intention of living anywhere but right here. And by that I don't mean Harlan County, I mean Ages. I've got lots of relatives here. I was born and raised here. It's my home. So we come back, and we done all right. In fact, I've made as much money and done here as well as I did over in Virginia. The opportunity was here, which it wasn't back in the fifties and the first part of the sixties, when the coal market was so bad.

I don't have any intentions of going back nowhere no more. Nowhere. Ages is my home. As long as I live, I'll be here. If there's anything to get by on.

10 MEANWHILE, BACK AT HEADQUARTERS

John L. Lewis is worshipped along Clover Fork. For forty years, he ruled the United Mine Workers with an iron fist. His craggy face, his leonine mane, his briar-thicket eyebrows, his rolling oratory in defiance of presidents, congresses, courts, and coal companies were the union personified. During the Bloody Harlan days of the 1930s and early 1940s, he played the union's hand—against a deck stacked by company money, company-owned sheriffs and judges, and unsympathetic governors—and he won. By the end of World War II nearly every miner in Harlan County was a UMWA member. The coal business was booming. Times were good.

But in the early 1950s the bottom dropped out of the coal market. Steam coal that sold for $4.25 a ton during the last days of the Truman administration dropped to $2.60 a ton during the early days of the Eisenhower administration. Meanwhile, wages were rising, and transporting coal out of the remote hollows of Harlan County was expensive. If a mine's coal was not of the highest quality and if it could not be mined easily and cheaply, it was no longer profitable. Mines closed by the dozens and operators packed up and left. So did 30,000 or more of the 74,000 people who had lived in Harlan County during the 1940s. With Lewis's acquiescence, many of the operators who remained began mechanizing their mines and hiring only a fraction of the number of men they had employed before. Hard times had come to Harlan County to stay, it seemed. Miners became more willing to work without a union contract, and negotiations toughened. The operators found the district officers and international leadership more and more willing to accept sweetheart deals, under which the companies would continue to pay their royalties into the union's Welfare and Retirement Fund, but were free to handle the men in the mines as they pleased. Such a contract was signed at Brookside in 1959.

Meanwhile, the miners lost control of their union. Lewis suspended the rule that district officers were to be elected by the membership and began appointing them himself. Cronyism became the rule. And the officers, who owed no allegiance to the men in the mines, became less and less concerned with conditions in the coalfields and more and more preoccupied with the millions of dollars in the Welfare and Retirement Fund, and with well-tailored suits and limousines.

In 1960, Lewis finally gave up the helm and was succeeded in the union presidency by Thomas Kennedy. In 1963, Kennedy retired and was succeeded by W. A. "Tony" Boyle, who had been a miner for a few years in Montana but had since led a much more prosperous career as a Lewis lackey. By then, a few voices within the union were crying for reform. So when the time came to prepare for the 1964 United Mine Workers convention in Miami, Boyle decided to have an unofficial police force on hand to keep order. To organize this force, he turned to an old crony, Albert Pass, secretary-treasurer of UMWA District 19.

District 19 encompasses a number of counties in Eastern Kentucky and Tennessee, including Harlan County. The men there have always been loyal to the union's leadership. In the early days, the miner's life had been harder there than in most places, and the operators' ruthlessness during the organizing wars had been unflagging. The union had been the miner's only friend in District 19, and he had never forgotten. Pass had no trouble recruiting.

One of the "White Hats," as the enforcers were nicknamed because of the white miner's hats they wore, was Junior Deaton. In a drawer at Junior's store is a photograph of Junior in his white hat, holding aloft a Boyle sign. Boyle, also wearing a white hat, has his arm around Junior and is grinning into the camera. He autographed the picture for Junior.

"Yeh, they called us the White Hats," Junior said.

They called us the wrecking crew, the goon squad, everything else. Really, though, it was like you was going to have a church over here, and you anticipated a disturbance, you know. You put some guards there to kind of keep it from getting out of hand. Well, at the convention, you've got about fifteen microphones. That's for delegates to use without having to walk too far. But, now, if anybody got out of line, why we was sort of manning those microphones. We had a fight. We had a battle that first day. And it was because of some anti-Boyle people just there to sort of wreck that convention. And we seen to it that they didn't. It looked pretty bad in the newspapers. They had some pretty good pictures of the fighting that took place. But it was peaceful from there on out.

Within a year, the union had deserted Brookside, the strikers were scattered to the four winds, and scabs were running the coal. Between the demise of the Brookside local and its resurrection nearly a decade later, a lot happened to John L.'s union.

On November 19, 1968, seventy-eight miners were killed by an explosion in a Consolidation Coal Company mine at Farmington, West Virginia. It was a UMWA mine, and some rank-and-file miners expressed concern that the Boyle administration seemed unwilling to fight for safer working conditions. In January 1969 a group of West Virginia miners organized the Black Lung Association to lobby for safer mines and more adequate compensation for pneumoconiosis and silicosis, the miners' work-induced lung diseases. One of the movement's leaders was a rank-and-file miner named Arnold Miller.

On May 29, 1969, Joseph Yablonski, called Jock by his fellow miners, announced he would oppose Boyle for the union presidency. He was a member of the UMWA power structure. As a former president of District 5 in western Pennsylvania, a member of the UMWA International Executive Board, and director of Labor's Non-Partisan League, the union's political arm, he knew where all the skeletons were buried. Boyle immediately fired him from the political job.

Yablonski's platform was a severe indictment of the union's power structure and a promise to return the UMWA to its rank-and-file members. It was a tough campaign. In Illinois, Yablonski was hospitalized for several days after he was struck down by a karate chop at a rally. He cancelled a rally at Wheelwright, Kentucky, after his life was threatened. He did not venture into Middlesboro, Kentucky, the headquarters of District 19, or Harlan, or any other part of Albert Pass's fiefdom. He knew Pass and he remembered where the White Hats came from.

He also knew that mine-closings and unsuccessful strikes had made District 19 a district of pensioners. More than half of the union's 5,000 members there were retired, and Boyle had told them that Yablonski would take away their pensions if he was elected. Simultaneously, Boyle raised the pension from $115 to $150 a month. On election day, December 9, 1969, fifty-two locals voted in District 19. The result: Boyle, 3,737; Yablonski, 88. The tally was reported so quickly that Yablonski immediately charged fraud.

Yablonski lost the election by a margin of almost two to one, but he refused to make peace with Boyle. He filed suit in federal court, accusing Boyle of buying the election with UMWA money.

On January 5, 1970, the bodies of Yablonski, his wife, and their twenty-five-year-old daughter were discovered in their home in Clarksville, Pennsylvania. They had been shot in their beds, and had been dead since December 31. Boyle said he was shocked at the news.

Subsequently, police and FBI investigations, interrogations, trials, and confessions uncovered a complicated murder plot that had been set in motion less than a month after Yablonski announced his candidacy, and was financed with union funds.

On June 23, 1969, Boyle told Pass and District 19 president William Turnblazer that Yablonski should be killed. Pass returned to Middlesboro and enlisted William Prater, a union field representative, to find someone to do the job. Prater contacted

Silous Huddleston, president of a UMWA local at La Follette, Tennessee. Huddleston called his daughter, Annette Gilly, who lived in Cleveland, and enlisted her help. Annette's husband, Paul Gilly, agreed to find someone to commit the murder.

The following September, Boyle and Pass met in Washington and worked out a scheme to raise money for the job. Pass wrote Boyle two letters requesting a total of $20,000 for "organizing expenses" in District 19. Boyle approved the requests. Then Pass and Turnblazer wrote out two sets of checks to twenty-three UMWA pensioners, who formed a phony "Research and Information Committee." The pensioners cashed their checks and turned over the cash to certain UMWA field representatives. The pensioners were told that they were kicking back the money to help influence local elections in Harlan and Bell counties.

Prater then passed on $6,000 of this money to Huddleston, who gave it to Gilly. Two would-be murderers, Claude Vealey and James Charles Phillips, had been hired by Gilly and were stalking Yablonski along his campaign trail through the coalfields.

As the campaign grew hotter, Boyle and Pass became more cautious. They decided that it was too dangerous to kill Yablonski before the election. The union might be suspected. So the word was passed down that the killers should wait. A few days after the election, another word was passed down. The murder plan was on again. After Christmas, Prater delivered $5,000 to Huddleston and promised another $5,000 if Yablonski was dead by New Year's Day. Phillips had dropped out of the murder squad, but had been replaced by Aubran Wayne Martin. On New Year's Eve, Martin, Gilly, and Vealey tanked up on whiskey courage, drove to the Yablonski home, and did the job.

Yablonski had noticed a suspicious car near his house, apparently staking him out. He had jotted down the license number, and police found it on a table by his bed. Three weeks after the murder, the gunmen were arrested. Vealey quickly confessed. Soon, Annette Gilly and her father were arrested. Boyle held a press conference and swore on a Bible that he was not involved.

Vealey pleaded guilty and turned state's evidence. Martin and Gilly were convicted and sentenced to death. In April, Annette

Gilly confessed, and said that her father, Huddleston, had told her the murder was approved by the "big man." "To me, that meant Tony Boyle, president, United Mine Workers," she said. A month later, Huddleston changed his plea from innocent to guilty and pointed a finger at Prater and Pass.

On May 1, 1972, a federal court overturned the 1969 election, declaring that Boyle had used union funds illegally to defeat Yablonski, and ordered that another election be held before January 1, 1973. The Miners for Democracy, which had been organized by former Yablonski supporters allied with the Black Lung Association, nominated Arnold Miller, Mike Trbovich, and Harry Patrick to oppose Boyle and his running-mates.

Boyle already was in the toils of the law. In March 1971, he had been indicted for embezzlement and illegal use of union funds for contributions to Hubert Humphrey's 1968 presidential campaign. A year later, he was convicted, sentenced to three years in prison, and fined $130,000. But he remained free on bond while he appealed. The now indignant rank-and-file abandoned him and flocked to Miller. The reformers won easily. They took office on December 22, and immediately opened the UMWA files to investigators.

In March 1973, Prater was brought to trial. Boyle testified on his behalf, but he was convicted and sentenced to life imprisonment. Prater confessed and named Pass as the man who ordered him to arrange for Yablonski's murder. In June, Pass was convicted, and was sentenced to three life terms.

Unlike the other conspirators, Pass did not confess. He would not implicate Boyle. Then, in mid-August, District 19 president William Turnblazer, who had always been a figurehead while Pass wielded the real power, broke down and told FBI agent Wallace Estill that Boyle had given the order to kill Yablonski. On September 6, he pleaded guilty to federal conspiracy charges. On the same day, FBI agents in Washington arrested Boyle. By the end of the month, he had been indicted and had attempted unsuccessfully to kill himself with a massive dose of sleeping pills.

In December, Boyle's appeal of his embezzlement conviction failed, and he was imprisoned. The following April 11, he was convicted of the Yablonski murders and was sentenced to life imprisonment. A month later, the star witness against him, William Turnblazer, was sentenced to fifteen years.

Two months after Boyle's conviction, the Brookside miners voted to ally themselves with the United Mine Workers. It was a bruised and battered union they were joining, but it was full of hope. Just two months before Arnold Miller took office, the federal courts finally decided a suit that had been filed in 1969, questioning the legality of suspending democracy in the districts. When Yablonski announced his candidacy, his district was one of the few left where the members still had the right to elect their own officers. All the rest had been put "in trust" and their officers chosen by Lewis, Kennedy, or Boyle and their lieutenants. The courts ruled this illegal and ordered new elections of district officers.

The decision fit hand-in-glove with Miller's promise to return the union to its members. He quickly instituted another reform, designed to prevent sweetheart contracts. No contract would be binding until it had been ratified by a majority of the miners who would work under it. Miners had not voted on a contract in fifty years. The ordinary miner now had a voice in his union's business. In District 19 and Harlan County, though, there were not many union miners left to speak. The UMWA had been driven out of mine after mine.

"Once Tony Boyle lost that election to Miller, I was as much for Miller as I was for even John L. Lewis," Junior Deaton said, "because I wanted the organization to survive. Another two or three years, and they would have declared this district dead, cause they wouldn't have been enough union mines running to keep a district here. But now that the Miller regime took over and done what they already showed me they're willing to do, I'm looking forward to a large district in this area here. I really am."

For Miller and the union leadership, more was at stake at Brookside than a UMWA contract for a middle-sized mine. As his first major organizing struggle in a weak and formerly corrupt district, the strike offered the new president the perfect opportunity to fulfill his promise of renewed UMWA militancy in the coalfields. The unprecedented $100-a-week strike benefit and Miller's frequent journeys to Harlan County gave the miners of Clover Fork more stamina and higher morale than they had enjoyed for a generation or more.

Even so, Duke Power and Eastover had resisted for more than a year, and the union's triennial negotiations for a new national contract with the National Bituminous Coal Association would begin in late August. If Miller could win big at Brookside, the union might regain the respect and allegiance of the thousands of miners who had defected during the hard years. But if he lost they would see the strike as just another verse of the same old song, and the operators, growing fat again on a new coal boom, might see him as a patsy.

The fight was in Bloody Harlan, bulwark of nineteenth-century industrial feudalism, and ground on which some of the union's most memorable battles of the past had been fought. It was in District 19, home of the White Hats, where Jock Yablonski had feared to campaign, where his murderers had plotted and recruited. If the UMWA and the reform administration could succeed in Harlan County, Kentucky, they could succeed anywhere. Brookside was the place to push all the chips to the middle of the table.

11 SPITTING & WHITTLING

Tuesday, the third day after the shooting of Lawrence Jones, dawned bright and cool on Clover Fork. It was jacket weather. It had rained during the night and the air was damp and clean. Large pools of murky water stood in the narrow strip between Kentucky 38 and the L&N tracks, where the grass had been obliterated by the feet of the pickets, and the Budweiser and RC Cola cans had been raked into piles. Water stood in the chuckholes of Brookside's mud lanes, too, and in shallow depressions along the paths from the back porches to the privies. On the clotheslines, washing that women had forgotten to gather in on Monday sagged like ranks of tired, soggy ghosts.

An uneasy calm had settled upon the villages. Eastover had not attempted to send miners into Highsplint since the beginning of Arnold Miller's five-day "memorial period" more than a week before. The few strikers now assigned to the Highsplint line just drove up the hollow before daylight, waited until past the time for the mine's first shift of workers to appear, then climbed into their cars and roared home to breakfast or to bed, or to the Brookside line to exchange early morning gossip with the pickets there.

Three men were pulling their assigned eight-hour duty by the railroad track. A half-dozen more, all with guns at their belts or handy in their cars, had joined them. Four sat along the track, like birds on a wire. Another lounged against a stack of railroad ties beside the roadbed. He was a young, big man, with red sideburns and huge muscles. The sleeves of his half-buttoned gray shirt had been cut away. His loafers were run down at the heels, and his orange jeans were grease-spotted and wrinkled, but his horn-handled revolver and its brown leather holster, slung low and tied snugly to his right leg, glowed from the care he had given them. He spoke in a high-pitched monotone, rarely pausing for breath.

"Look at them dogs over there." He pointed toward a pack of seven or eight mongrels cavorting dog-fashion near the Eastover office across the highway. "Ain't they having fun? They's really after that little old bitch, ain't they? One of them's mine. He'll get his share. You wait and see. He's a tough old dog. Uglier than hell. Good watchdog, though. Anybody comes anywhere *near* the house, he bristles up and growls. I been sitting in the front room holding my shotgun in my lap these nights. Lot of shooting been going on up around my place. Lot of scabs lives up there. They ain't getting *me*, though, I'll tell you. Not without I take some of them with me."

The others paid him no mind. Tub Rainey, sitting in his pickup, his legs hanging out the open door, moved his knife blade slowly along a cedar log, watching the pink shaving curl like a clock-spring, then drop. A week ago, when he began, the log was as big around as his arm. Now it was as slender as his finger. On the pickup radio, Jeannie C. Riley wailed of love and infidelity. A young miner pulled up in a shiny new muscle car with a loud muffler, gunned the engine, then cut the ignition. He sat for a few minutes, staring through the windshield at the others, then decided not to get out. He folded himself up on the front seat and went to sleep.

Over on the tracks, two pensioners in baseball caps chewed and spat and talked politics. "Now, that Marlow Cook and that Wendell Ford, there's a race where we just can't *help* but get screwed. I ain't voting for neither one of them. They's both against us. They ain't got no *right* to be a senator. Hell, I ain't even voted for Happy Chandler since that time he sent the National Guards in on us."

Tub turned up the radio. Clover Fork's most popular song was playing.

I'm from Harlan County, Kentucky,
And, son, consider yourself damn lucky
You're not already six feet underground.

Silence fell as the Harlan newscaster began his drone through the doings of Henry Kissinger and Gerald Ford. He finally got to the story the men were waiting to hear, then disappointed them, repeating words they had heard a dozen times or more, about their picket line Monday morning at Highsplint. Officials of Duke Power and the union were scheduled to meet with the federal mediator on Wednesday in Washington, he said.

"Shit, don't that feller ever get no new news?" one of the pensioners asked. "He's been saying that same damn thing for two days now. Never changes ary a word."

"Well, as long as they don't try to work them Highsplint men, they ain't nothing new to tell," Louie Stacy said. "I don't want to hear nothing new till they sign that United Mine Workers national contract. That's the only news I want to hear. Norman Yarborough's name on that dotted line."

Tub grinned. "You and me, buddy. He might as well go ahead and sign it now, cause he's gonna *have* to someday."

The young man with the tied-down holster divined great significance from the fact that the newscaster had said the meeting would be attended by Carl Horn, president of Duke Power, and had not mentioned Yarborough, president of Eastover, who had been the company's sole spokesman throughout the strike.

"Old Carl's taking over the negotiations," he said. "He ain't even letting old Norman go to the meeting. He's saying, 'Norman, old buddy, you done fell over in your own shit.' "

"Maybe," one of the pensioners replied. "But I ain't gonna let my hopes get up. I'm expecting the worst and hoping for the best. I been let down too many times."

"I'm going down to Junior's and see if he's heard anything," Louie said. He crawled into his Volkswagen and drove around the curve to the little concrete-block building with two Texaco pumps in front and crates of empty soda-pop bottles stacked high against the wall.

Crammed into the small room is everything a miner's family is likely to want on short notice. "If I don't sell it, you don't really need it," Junior likes to say. Bread and cakes and butter, meat and green beans and peanut butter, chili and condoms and cough-syrup, snuff and cigarettes and chewing tobacco, sun bonnets and pocketknives and stovepipe, dolls and bubble gum and lollipops, stove shovels and roach powder and car tags proclaiming, "I'm a Coal Miner's Son." It is all there somewhere, and Junior can find it for you. Taped to the open front door are bits of information and entertainment. Clippings from *The Courier-Journal*, which the milk routeman brings every morning but Wednesday. News pictures of President Ford and Henry Kissinger, with comic-strip balloons ballpointed in. "After you finish with the Arabs, look into that Brookside strike," Ford tells Kissinger. "Let Rocky buy it," Kissinger replies. Sometimes there is a poem about the strike or Richard Nixon, written by Junior and painstakingly typed in capital letters by his daughter. If a stranger stops and wants to know where he is or where he is headed, Junior can tell him. He will also ask the stranger where he came from, and what he is doing in Ages. He might take the stranger's picture with his Polaroid camera, and say, "Next time somebody tells you, 'I ain't seen you in Ages,' you can say, 'Well, that's because you wasn't there.' " If the stranger has a low tire, Junior will lend him his tire gauge and tell him where the air hose is. If Junior likes the stranger and it is a slow day, he might take him to the back room, where there is a set of drums and an electric organ and a baby grand piano, and play some boogie for him. Or he might turn on the stereo and put on the tape called "Rocking in Ages" that he recorded with his son.

Junior is a short, dark-haired man, thickening and rounding into middle age. His rough face is solemn most of the time, except for his brown eyes, which glint with irony and wry humor. During the strike's cold, dull winter, when the mountains were gray and ugly, he tried his hand at poetry, and decided it was fun to make things rhyme. "I swear, I have more *fun* doing that," he said. "I get tickled at myself. In the wintertime, especially, when I close the store, first thing I do is go home and read the *Knoxville News-Sentinel*. I read the Louisville paper in the morning and the Knoxville paper at night. And after I read it, I ain't got nothing to do. Just watch television. And I don't like that unless it's a news program or something like that, you know. So I begin to write poetry, and I done that all winter. I wrote some that would tickle you to death. It's just to make every other line rhyme, but you can really have some dillies, especially on Watergate. After I put them on the front door for a few days, I take them to my mother. She likes them."

Junior is one of the best-informed and deepest-thinking residents of Clover Fork, and if Ages had a mayor, he would be it. He helps miners fill out their income-tax returns. He knows influential lawyers in Harlan. When politicians need votes on Clover Fork, they seek his endorsement. He knows who to see and what to do to get state funds for a children's playground at the Ages school.

"I like to know what's going on," he said. "I read three newspapers a day, counting that scab paper in Harlan. That's the first thing I do when I go to another town, is get a newspaper. No matter where I go. It's a motel, hotel, or somebody's house, I manage to get a newspaper. I'm a newspaper hound, I guess."

He was in the mines for sixteen years, and was involved in the Brookside strikes of 1959 and 1964. Like Louie and the other staunch union men, he left Clover Fork when the union collapsed, and went to work for a UMWA mine in West Virginia. His hip broke when the shuttle-car he was driving jumped the tracks and crashed into the rock. "I laid in the hospital for seventy-two days. That's a long time to stay in a hospital. And I said, 'Well, I guess I'll just quit. Quit the mines.' I need four more years for a pension. I'll tell you, it's a very possibility that in a couple of years, maybe when this thing is settled down here, that I might want to venture in down there and pick up those four years. I'd like to do it at home, you know. *If* I do it."

When a union man has time on his hands, he wanders in to talk to Junior. If Junior is too busy, the man will take a seat on a crate of canned goods by the stove, or on the bench outside the door, and drink a soda pop until another buddy wanders in.

"Carl Horn's going to Washington and Norman Yarborough ain't," Louie said. "What do you reckon that means, Junior?"

"Don't know. It might be good."

"And it might not, too."

When the afternoon turned steamy, the pickets moved across the highway to the shade of a line of trees bordering company property. The company had strung a thick, greasy steel cable from tree to tree to mark the boundary between management and union territory. Over a sheet of newspaper spread on the ground, three strikers gambled for nickels and dimes at a card game called Tonk. Another striker stood in the middle of the highway, holding a plastic Clorox bottle with the top cut off. A union sympathizer stopped, pulled a long black change purse out of the bib pocket of his overalls, and dropped a few small coins into the jug.

"What you fellers going to do with all that money?" he asked.

"We're gonna buy some bullets to use on the scabs," the striker replied.

"Well, hit me again after the first of the month. I'll have more money then."

A slender man wearing sunglasses watched the little pink shaving curl up on his knife blade and float to the ground. He glanced up at Louie, stretched out on the hood of a car, leaning on his elbow. "When you gonna write us some more signs, Louie?" he asked. "These trees look awful bare."

Louie smiled slightly and shrugged. "It's too late," he said. "Now is either settlement time or shooting time."

12 THE TOM PAINE OF CLOVER FORK

If Louie had been born in another time and place, or if he had not had to quit school at thirteen, he might have been a pamphleteer or an editorial writer. His style is more direct and defiant than most publishers in a slick, urbane time would permit on their pages, though. It harks back to days when men got really angry in print. If he had lived in Boston during the last quarter of the eighteenth century, he might have put a little more bite into the broadsides cranked out by Sam Adams and Paul Revere and Tom Paine.

In conversation Louie is soft spoken. His anger smoulders quietly behind his light blue eyes. But during those winter nights when Junior Deaton was struggling with his rhymes, Louie found *his* literary voice, too. Kneeling on the kitchen linoleum with paint can and brush, he lettered his defiance onto large sheets of tin, carefully stabbing dots between the words:

WE · WILL · NOT · EXCEPT · A
SWEET · HEART · CONTRACT
NEVER · NEVER · NEVER · NEVER
WE · ARE · HERE · TO · STAY
WE · WILL · BE · HERE · ON · THE
PICKET · LINE · UNTIL · CHRISTMAS
1975 · IF · NESSARY · TO
WIN · THIS · STRIKE · AND · A
UMWA · CONTRACT
WE · WILL · NOT · BE · BULL · DOGED
OFF · OF · OUR · PICKET · LINES · BY
THE · STATE · POLICE · OR · THAT
COAL · MINE · OPERATOR · BYRD

HOGG · OR · THAT · HARLAN
COAL · ASSOCIATION · OR
ENYONE · ELSE · WE · BE · HERE
WHEN · THE · MORNING · COMES

He carried his signs to the picket line and nailed them to a tree. Photographs of them published in newspapers and magazines made them symbols of the strike and the determination of the miners.

By spring, some company man was sick of looking at them. One morning the pickets arrived at their station and the signs were gone. When Louie found some more metal sheets, he went to work again:

PUT YOUR [picture of heart] in U.M.W.A.
OR GET YOUR [picture of donkey's rear end] OUT
I PREFER A UNION MAN
FOR A FRIEND. HOW ABOUT
JOINING THE U.M.W.A. AND
BE A GOOD FRIEND
SACRIFICE A LITTLE TODAY
TO HAVE A GOOD SAFETY
PROGRAM TOMORROW
BROOKSIDE WILL NOT BE
SCABED ALL US UNION MEN
OUR WIFES CHILDREN LABOR
BOARD RETIREES C.B.S. AND THE
WHOLE U.S.A. HAS A BIG
[picture of eye with arrow pointing to it]
ON EASTOVER
AND WE ARE NOT ASLEEP
I DID NOT HAFT TO USE
SOUTHERN LABOR TODAY
CHANCES ARE I WONT HAFT
TO USE IT TOMORROW (Pu Pu)
SIGNED LOUIE STACY

This time, Louie decided to add a postscript, presenting Norman Yarborough's view of the strike, as perceived by the miners.

"I just had one more piece of tin," he said, "and I had a lot to get on it. I had to write small. But I didn't have no paint-brush little enough to do it. So I just got me a match and stuck it in my mouth and frizzled out the end of it with my teeth and used that."

My opinion this is
Norman Yarboroughs view. Let Me Make This Perfectly
clear i am not a crook. i Am Not Getting Big Payoffs
From the Coal Association. i Did Not Give Judge
Byrd Hogg Thousands of Dollars to Put You
People in Jail. Boy Do i Hate A Safety Program i Rather
Have You Boys Crippled and Maimed Up i Did Not Call
Cov. Ford To Have A Lot of Extra State Police
Sent in Here To Bulldog You All Off of your
Picket Lines. I am Not Playing Duke Power Co.
For A Sucker (Not Much i Aint)
Ho Ho. My Good Brother Do You Believe The Above Crap?

In April, CBS telecast some film of the Brookside camp, comparing living conditions there with conditions in the Harlan County coal camps of forty years ago. CBS concluded that coal camps have not changed much. A few days later, the *Harlan Daily Enterprise* published an editorial lambasting "foreign" newsmen who came to Harlan County to cover the strike and presented unsavory pictures and accounts of life there. The *Enterprise* declared that the living conditions that prevailed during the 1930s have all but disappeared from the county.

"That really burned me up," Louie said. "They were trying to pretend that there ain't no mud streets left in Brookside. Ruby and me set down and started writing a letter to the feller that runs the paper. It took us all night and sixteen sheets of paper. We gave them a good blasting."

"Another thing," the letter said after Louie had warmed up for a couple of paragraphs, "your editorial sounded like you don't believe that living conditions like Brookside's that were shown on TV don't exist anymore.

Well, they do. They got outside toilets and still carry water from a spigot eight or ten houses away. They got mud streets and take a bath in a wash tub just like back in the '30s.

The old houses are not insulated and it takes 12 to 14 tons of coal to heat the four-room houses for a winter just like back in the '30s. . . . All the pictures shown on TV were taken here at Brookside the week before they went on the air so how can anyone say those conditions don't exist anymore? . . .

I admit that all of Harlan County is not like Brookside. It looks like now they are starting to do something about living conditions here. They are tearing all the houses down. But the people have nowhere to move to. A while back there was some talk of a housing project for the miners. But I will have to see it to believe it. This company don't care enough for its people to provide them with decent housing. All they want is to gut these mountains and leave. . . .

Don't just report one-sided news and don't run your paper for a few bigshots in Harlan. Remember they just buy a few papers. We the people in these coal camps and up these hollows buy most of your papers. Be your own man. As for myself, I am my own man. Eastover does not own me. Or the UMWA does not own me. . . .

At the very mention of gas, oil or milk, the sweat just pours from Nixon's upper lip. At the very mention of a safety program Carl Horn and Norman Yarborough start to sweat all over. Nixon hates all coal miners. Yarborough only hates UMWA miners. . . . All we want is a safety program. And for the men to be treated halfway decent. We are not asking for the moon, just a small slice of it. . . .

Ruby had bet Louie five dollars that the *Enterprise* would not print the letter, but it did.

"She ain't paid me, neither," he said. "Guess I'll just have to take it out in trade."

Ruby giggled. "Why do you go on like that?" she asked.

13 THE DEATH OF A BROTHER

Tuesday, Lawrence Jones passed his twenty-fourth birthday. Wednesday, about five o'clock in the evening, the doctors pronounced him dead and turned off the machines.

The strikers had known for four days that Lawrence was going to die. They could not really believe in the miracle that the wounded man's wife and mother had been praying for. They were not sure it would be good for Lawrence to survive. What kind of life can a man live with part of his brain gone? He would be nothing but a vegetable, that's all, they had said.

But when news of his death spread up Clover Fork, none of that mattered. One after another, the miners received the word, felt their stomachs contract in sudden sickness, then felt the sickness turn to rage and spread upward in a great rush of blood to their brains. They counted their ammunition and cried for vengeance, and Mickey Messer called a meeting of the union.

The crowd arrived at the Evarts Multipurpose Center armed to the teeth and in a lynch mood. The soft, angry conversations were of eye for eye, tooth for tooth, blood for blood. They milled, glaring into each other's eyes, as they were told what they already knew: "Our brother, Lawrence Jones, is dead."

A union organizer, Houston Elmore, quickly gave them the one piece of news that might hold back the tide of anger. Norman Yarborough had been called to Washington. His plane was in the air at that very moment. Serious negotiations were in progress. Settlement might come before the morning.

The Reverend Otis King, called Tag by his neighbors at Verda, stood among the people, his arms uplifted, his gold-rimmed glasses flashing in the glare of a camera crew's lights. His hoarse, rasping voice cried out: "We've suffered so much, folks, and we're so near the end! Let's don't ruin it now! Let's hold our peace for one more day! Just one more day!"

Anyone who wished could remain at the center to await the telephone call from Washington, Mickey said. Then he adjourned the meeting.

Most walked out into the night, to put their children to bed. The others gathered into small knots in the gloom of the large room, wondering how long their wait would be.

14 OF GOD & THE UNION

The Reverend Tag King is an old man with large, rough hands. He lives in a white frame house at Verda, not far from where Jones Creek flows into Clover Fork. Just up the hollow is the community of Jones Creek, home of the families of Lawrence Jones and Billy Bruner. The Reverend Mr. King has a large yard with many tall trees, and a large garden. He and his wife keep a milch cow and chickens and cats and a dog and a great-grandson they have adopted. The Reverend Mr. King's pulpit, the Turner Baptist Church, stands on a rise just up the hollow from the house.

The preacher is seventy-two years old and has lived along Clover Fork all his life. He worked forty-six years, from 1918 to 1964, in the mines. Bloody Harlan is a vivid memory, but the strikes, the beatings, the battles, the killings have merged into one long nightmare. On April 2, 1941, King and several other unionists on an organizing campaign walked into the company store at Crummies Creek, south of Harlan Town. A deputy sheriff was there, stationed behind a machine-gun emplacement on the meat counter. As the miners walked through the door, the deputy opened fire. He killed four men and wounded five. That is one account. As always, there are others. The bloodbath became known as the Battle of Crummies Creek. King no longer remembers how many men died there, but he remembers their dying. He does not remember what year they died, or which strike it was. It could have been any of them.

He leaned back into the sofa, scratched his thinning gray hair, and said:

I was borned up here on Big Black Mountain. I come down here and married me a wife. We moved away for a couple of years at a time sometimes, but we come right back. We been married fifty-two years. We had a big blowout up at the center. My boy—he lived in Ohio then; lives right down here now—him and my daughter come from Pennsylvania, and we went to the center, and they give us a big party. Fiftieth wedding anniversary. Golden wedding anniversary is what you call that, ain't it?

Let's see, I believe I was about thirteen, fourteen years old when I went into the mines. No, let's see. Maybe I was sixteen. I went in early. I had to tell a lie about my age. But I needed to work. It was back yonder when we didn't have nothing. Just what we could dig out of these mountains and get out of the trees and what we could farm on these mountainsides was all we had to live on. And when these public works come in here, mines and railroads and stuff, all us young fellers got on that quick as we could.

I first went to work on the railroad, on what they call grading. You know, back then, they just had old mules and scrappers. I helped build that railroad up Bailey's Creek holler, and then I helped build that one up toward Black Mountain. And then I worked after the steam shovel up Black Mountain when they's opening up the coal mines.

So when the mines started at Black Mountain, they organized the union. We had an awful time with that, and we had a strike then. Way long years ago. Peabody Coal Company. I don't remember what year it was, but we had the awfullest strike you ever saw, nearly. You know, we'd work and wouldn't get nothing for our work, and we couldn't hardly live, even working every day. And we just organized a union. There wasn't no big checks or nothing then, you know. You had a strike and you was on your own. There wasn't no handouts. We just done the best we could.

Well, back in the thirties, why I was working around here, and we was going in the mines about six o'clock in the morning, and we didn't know what time we'd get out that night. Whenever we'd get our place cleaned up. They didn't care when we got it cleaned up. Sometimes it'd be ten o'clock, eleven, and occasionally it'd be twelve o'clock in the night. We was working every day, and sometimes we'd go over there to the company store and try to get them to advance us something on our scrip, so we could buy something to eat. They wouldn't never give us but about half of what we asked for. They said that was all we was entitled to. We'd used up the rest, they said, buying our oil for our lamps and our powder and tools and stuff. Yeh, we had to furnish all that ourselves. We had one of the awfullest times there ever was.

And we was organizing, and people was killed, and they sent the National Guard, the State Militia, whatever you call them, in here on us, and they was shooting, and they killed some people. And then we had a bloody battle at Crummies Creek, and I don't know how many people there was killed. They's killed within fifteen or twenty feet of me. They just happened to not hit me, some way or other. We went to organize that place, and they had machine guns ready for us when we got there. So when we got in the building, they just opened up on us. So much shooting. . . . They was shooting us down like dogs.

Then we met right here in Verda one evening, and that got us a contract, I guess. They was about ten thousand people right over there in Verda, and the line reached pretty near up from Evarts down to Ages. The National Guards was there, and they was treating us so bad, and we was just down to where there wasn't nothing. Five had starved to death. And them union fellers, they just went right in and went to fighting them National Guards. And then some of them come off the highway into a wire fence, and we was all there. The childrens was there, and the wives was there, and just about everybody in Verda was right there. Them National Guards, they had them machine guns, and they had their sand bags on them, and they was pointing them around thisaway and thataway. If they'd started shooting, I guess there'd been maybe a thousand people killed, in spite of all that could be done. And we win that contract finally, and we went to organizing, and we organized this county just about solid union.

John L. Lewis was a great leader. He's one of the greatest. But when Tony Boyle got in John L.'s place, it wasn't long till the union went to going down. They wasn't no activity, and they wasn't nobody doing nothing. So, first thing we know, we're almost down to where we started from. But I think we got another great leader now. I do have a lot of confidence in Arnold Miller.

Now we had a boy killed. I've knowed Lawrence ever since he was a little boy. He was a little related to me. Me and him was kin. And I had a special friend—he's a good Christian man—shot once over the heart and once under the heart. Minard Turner. But he come out of it right quick. He's a wonderful man. And he was shot.

Course, we've been trying to organize, and we been on the picket line, but we're doing what we believe is right. We're having to patch up what we've lost of what we won back in the nineteen-and-thirties. We're having to reorganize all over again. I don't know what happened, but it looked like the union leaders just went off and set down somewhere and forgot all about us. That's the way it looks to me. And all that killing up in Pennsylvania? Why, that started a great row, and they wasn't no use in that. They wasn't no use in killing the Blonskis, or however you say that. They wasn't no use in them people being killed like that.

They wasn't no use for this boy to be killed. He wasn't on the picket line. He was up there in the holler about a mile. And they wasn't no use of that. They wasn't no need of Minard getting shot. He just walked up to the car and asked them to honor the picket line, and that feller just out and shot him. Didn't ask no questions or nothing about it. Just opened up on him and shot him.

It looks like what poor people gets, we always have to fight for it, and be killed and beat up and treated awful bad. The coal operators has millions and millions of dollars laid back, but they're ready to hire fellers to kill us, or break strikes, or just do anything to us. Instead of signing that contract, so men can have money and work and raise their families and educate their children, they'd rather keep us down, where we have to beg to them, you see. And all the time, they're laying the money back. Millions. If somebody kills some of us, they're ready to buy them out of it.

I'll tell you another thing. Right here in Harlan Town, and most anywhere you go, all the law's against us. Wendell Ford sent the State Police in here. They was about seventy-five State Police here one day, and them women laid down on the ground, and the police grabbed them up and drug them and pitched them in the car and took them to jail.

I was there. I was preaching a revival meeting there aside the road. The judge had said we couldn't go there to picket, so I just decided to have a revival there. I preached to anywhere from seventy-five to three hundred people there, through a loudspeaker. And I told them police on that loudspeaker not to do that no more. "They's enough of you here to pick them up and put them in the car," I said. "Don't drag them all over this highway and all these gravels and stuff. That ain't right." And they told somebody to tell me that they wouldn't drag them no more.

I've always been active in the union, and I will be right on. I'll always be a hundred-percent union man. The operators made a union man out of me, the way they treated me. They made me what I am, and I'm glad of it.

We need a contract. That's what the boy died for. That's what Minard was shot for. That's what the women was drug around and their children put in jail for. And the company ought to want the men to have a good living, and they ought to want the children to be fed and clothed and educated and all. But they don't see it that way. They'd rather lay it all back, and break the strike, and keep the coal miners down to where they can't have much to say, so all they can do is go on and work with bad top and dangerous conditions, and if men don't like it, the boss can say, "Well, get your bucket and go home." That's what they tell you. That's what they told them at Brookside.

People just don't know what men has to go through in

these mines, working with hanging rock just ready to come down on you. I had two fingers cut off. I had this ankle broke. I had my back broke, and I got arthritis of the spine. The only time the operators has any use for men like me is when we can work. When we can't work, that's it. You get out.

Before we organized, they used to work us night and day. Sometimes I'd come in and I wouldn't even take a bath, it'd be so late. I'd lay down behind the stove and sleep two or three hours while my wife was getting breakfast and ranging around, and then I'd get up and eat and go back to the mines. I come to the house only two times one week, and stayed only a hour or two each time. I laid down on the porch with my bank clothes still on, and I went for a whole week in the mines. My little boy would say, "Where's Dad at? I ain't seen him in a long time." And my wife would say, "Honey, he goes to work before you get up, and when he comes in, you're asleep." They was going to school, and they wouldn't see me from one Sunday morning till the next.

A coal miner just lives one day. Whatever they make, they spend. I didn't have time to figure no future or nothing. All I thought of was going back to work tomorrow. That's all the future. Just try to make it today, and go back tomorrow and try again. On Sunday, I got drunk. I didn't feel like doing nothing else. I was an awful drunk. I'd drink anything. I wasn't no Christian then. I was saved in '41. The eleventh day of September, nineteen hundred and forty-one, I was saved. The church where me and my wife got saved at is tore down and moved now.

On June the eleventh of '44 I was ordained as a minister of the Gospel, and on the fifteenth day of August I took a church. And I'd go to the head of these hollers and have prayer meetings with the people night after night. Oh, I was strong then, and always in good health. Up yonder where Lawrence lived, I'd go up there maybe three or four nights a week. We'd pray and read the Bible. That was after we organized the union. We had time then to come out and go to church and do things. We was getting along awful successful in our church. We had it full of people, and we was going all out. And about every Sunday morning, I'd be abaptising somebody somewhere.

But in all my work, and all my ministry, I never did get away from how I was treated. I've worked in mines where there's such a little bit of air that you might have to strike ten matches to light a cigarette. I seed a feller hold a handful of matches in his hand and struck them all at the same time, and he never got to light a cigarette off of all them matches. You just don't forget things like that.

But I helped pave the way to get the coal mines where it is now, and I'm proud of that. The hard times just made a good union man out of me, and I'll always be one. It's just in me. I'm proud of being a United Mine Worker.

You know, the Lord is first. His saving grace is first. I don't let nothing get between me and my church and my Lord's work. Nothing. And then my Mine Workers is next. It's second. That's the way I feel. That's my conscience. And, you know, I feel like I've got a clear conscience. I feel like I've been fighting for something great.

15 GUNFIRE & SWEETHEART WORRY

In the small hours of Thursday, gunfire broke out on Clover Fork, and women and children jumped from their beds and hugged the floor. The gunfire was rapid and sustained, and it came from many guns. The old ones thought it must be Evarts or Crummies Creek all over again.

This time, though, the guns were fired in celebration. A few minutes after three o'clock, the word had come from Washington. Norman Yarborough's signature was on the dotted line. The miners still waiting at the Evarts Multipurpose Center burst out of the building, leaped into their cars and trucks, and took to the roads, their horns blaring through the hollows, their voices whooping the news, their guns emptying carefully stockpiled ammunition into the black sky. At Jones Creek and Kenvir and Kildav and Verda and Evarts and Brookside, awakened families threw their bedclothes aside and joined the celebration. But it was brief. Most families just thanked God or Arnold Miller and went back to bed. Some drank a cup of coffee with their neighbors and talked awhile first.

Louie Stacy's trailer is away from the road, and he was unaware of the settlement until Darrell Deaton phoned him, after four o'clock. He dressed quickly and drove up Kentucky 38, looking for the party. No cars were on the highway. Deaton's Grocery was dark. A voice hailed him from across the highway, from a lighted house near the Ages post office. Louie pulled off the highway and drove carefully through a deep puddle up to the porch.

A half dozen men were there, lounging about the porch or leaning against a car parked in front of the steps, drinking beer.

One, his pistol still strapped to his hip, had been drinking longer or faster than the others.

"*Wooo!*" he yelped. "Anybody know what time it is?"

"About five o'clock, ain't it?" someone replied.

"I been up twenty-four hours right now. *Wooo!* The old lady'll *faint* when I tell her about it. I ain't shut my eyes in twenty-four hours now!"

He danced a little dance in the yard, his holster flapping against his side.

"Twenty-four hours!" one of his companions shouted. "Give that boy a goddamn medal!"

The tipsy one stopped dancing and pointed a finger at the speaker. "That boy cusses something awful," he said. "He uses profananity all the time." He tried again. "Profanity."

"Goddamn. What's so great about being up twenty-four hours, buddy?"

The man ignored the question. "Hey, listen, you got any shells left? Can't anybody sell me a box of shells right now? I want to shoot some more. *Wooo!* One more time! *Wooo!* I wish I had a thirty-ought-six and two boxes of shells right now. I really do. I done shot up sixty-seven shells tonight."

"You look like Wyatt Earp, buddy, with that big old gun hanging on you there."

"I'm out of shells for it, though. You got any, buddy?"

"You'd just shoot yourself in the foot. You're drunk, buddy. Go soak your head in that mudhole over there."

The man peered at the mudhole. "I'm going home," he said. "And I'm gonna kick the door down."

"Yeh, go home. You been up twenty-four hours."

"I'll tell you one thing." He wagged his finger at the amused loungers. "I'm gonna wake up the old lady before I go to sleep. I sure am. And she's gonna say, 'You sorry son of a bitch, where you been?' And I'm gonna say, 'Shut up. We got the contract.' *Wooo!*"

"Yeh. Lay it on her, buddy."

"Now don't go encouraging him," another celebrant cautioned. "You're liable to get his ass whupped."

"*Wooo!* One more time! They signed the goddamn contract! I was there when they *done* it. The man on the phone, talking to Washington, he said, 'Let me talk to the chief organizer hisself, Tom Pysell.' And the guy said, 'This *is* Tom Pysell. They just signed the contract.' And the man just laid down the goddamn phone and said, 'Hey, they just signed the *contract!*' I was there."

"We oughta have the biggest goddamn party this county ever seen," one of the men said.

"We gotta have a funeral first," Louie replied.

"Yeh, that's the bad part about it. Sticking a twenty-three-year-old feller in the ground. That's the bad part about it."

"But them scabs is on their way out now," Louie said. "All over Harlan County."

"Their mothers is setting there crying like a baby right now."

"Everything will be all right, baby," said the man with the gun. "Everything will be fine, baby."

"Hey, how long we been out now?"

"What's the day of the month?" Louie asked. "Twenty-eighth? We been out thirteen months and two days. Naw, it's the twenty-ninth now, I guess. But I guess today don't count, since the ink is on the contract."

"Been a lo-o-ng time. But we got no worries now."

One by one, the men drifted to their cars and drove away.

Louie sipped his coffee from only one cup. There was no need to hurry now. He drank silently, a gleam of joy in his eyes. At six o'clock, the Harlan radio station came on the air and announced the end of the strike. Louie listened, smiling. Then Eastover's public relations man came on the air with a victory statement. A no-strike clause had been the cornerstone of the company's position throughout the negotiations, he said, and the company was pleased that the new contract contained a no-strike clause. Having won that, the company was glad to sign with the United Mine Workers.

Louie's happiness turned to disbelief. "What was that? What did he just say?"

The public relations man continued the long statement, emphasizing the company's satisfaction with the contract.

Louie was angry now. "We've been sweethearted again! It sure sounds like it! What the hell have we stayed out for thirteen months for, anyway, if we're gonna give them a damn no-strike clause? That's why we stayed out, ain't it?"

At seven o'clock, Louie struck out up Kentucky 38 again. Miners already were congregating at Deaton's Grocery. Those who had not heard the broadcast were still jubilant over the settlement. Those who had heard it were troubled.

"They've sweethearted us, buddy! Don't you understand that? If we got a no-strike clause, we ain't got nothing!"

"I don't believe that. Arnold Miller wouldn't do that. Somebody's lying to us."

"It was on the radio, buddy. I heard it myself. Ain't that right, Louie?"

Louie nodded. "That's what it said. If that's what we got, I ain't going back. I'll leave first. I'm tired of running every time we have a strike, but I'll leave first."

"You and me, buddy. I won't go back. *None* of us will go back. We'll stay out till we're all gray-headed."

The men moved down the highway to the stack of railroad ties beside the track. A few miners already had gathered there. The guns were absent now, but the men milled nervously, their voices full of doubt. No one knew for sure whether they were still on strike. They did not know whether they had come together to celebrate or to picket.

Suddenly, Carl Noe, Tub Rainey, and two others bolted for Carl's pickup and sped down Kentucky 38 toward Harlan. "They're going to talk to the organizers," someone said. "We gotta know."

The newscast came on again. The man sitting in the car turned up the radio. Eastover's statement was repeated. The men peered at each other. "It sure looks like it," one said. "Looks like we been screwed again."

"Not me, buddy," Louie said. "I ain't going back if there's a no-strike clause in there."

"You and me, buddy. None of us will go back."

Thirty minutes later, Carl and his companions returned, jubilant again. Carl sprang from the pickup. "Old Norman's bullshitting us," he shouted. "The organizers called Washington, and we talked to Tom Pysell hisself. Norman signed the goddamn national contract, and that's all there is to it. There ain't no no-strike clause. They're going to buy an ad on the radio and tell people not to pay no attention to what Norman's saying."

The men relaxed into jokes and laughter. They compared potbellies they had grown during the long months of inactivity. Carl lifted his white T-shirt and peered at the huge hemisphere of flesh that bulged over his belt. Suddenly he sprinted down the highway toward the company office, holding his shirt above his bouncing belly.

"Arnold Miller paid to put this on me, Norman!" he shouted. "But, by God, you're gonna pay to take it off!"

Louie was ready to take it easy. He put a tape of hard Jeannie C. Riley honkytonk music on the stereo and leaned back into the big chair, grinning. He was in a talking mood. "This is the kind of music I used to like, back when I was helling around," he said.

I used to drink a lot, hang around dance halls, get in fights. I remember one time I was drunker than a skunk, and I started to walk across the dance floor, and just keeled over and passed out, right in the middle of the floor. People was dancing all around me. About the time I passed out, a couple of fellers got into a fight. Pretty soon, everybody in the place was fighting, all over the dance floor. Later, somebody told me people was real careful not to step on me while they was fighting, trying not to hurt me. Wasn't that nice? They'd just come up to me and stop, and then go around me real careful to get to whoever it was they wanted to hit.

A man's gotta settle down sometime, though. I ain't had a drink in a long time. A man can waste a lot of time drinking. A lot of money, too. A man's gotta start taking care of hisself when he gets older.

I've worked in the mines too long. When we first went out on strike, I probably could have went and signed up for my rock dust pay and my Social Security. If a man gets down bad, he can get that anytime. But he can't get his United Mine Workers pension till he's fifty-five. I've got rock dust. I know that. When I lay down at night, my chest feels like somebody come along and set a big heavy plate of food on top of it. My chest has felt a lot better since we been out on strike. Not so heavy, you know. I guess it'll be clogging up again now.

If a man's not in too bad a shape, he don't want to retire too young. But I don't want to wait till I'm too old, either. I mean to enjoy life. We've had some miners, they just refuse to give up. They work till they're sixty-five or seventy years old. Well, then, by the time they quit, they're so near dead they don't enjoy life. I mean, they ain't got time to enjoy it. About the time they get their pension, they're dead. They're already sticking them in the box.

A coal miner when he retires, if he don't get him some kind of hobby, he don't live long. I guess he's been used to working all his life, and he's got too much free time, and he worries too much, I guess. Even if they retire young, like I'm going to. I'll be fifty-five when I get my pension. They don't live long.

When I retire, what I'm gonna do, if I've got the money, is make a movie or something like that. Maybe go around and cover strikes and things, making movies of them, just for my own personal use, you know. Try to get me about a four-thousand-dollar outfit and just make movies. Now that would be a good hobby to take up, wouldn't it? It would keep you occupied. And I like music. Maybe I could record some music. And I'm hoping to get out someplace with about two or three acres of land to it. You know, where a man can raise a garden and get away from where it's so heavily populated and maybe raise us some hogs and chickens.

He took a long pull on his coffee and listened to Jeannie C.

belt out the honkytonk songs. His eyes roved over his comfortable trailer living room. Ornamental clocks that do not run were displayed on the walls and the top of the television set, among numerous gold-framed photographs of his daughters, his grandchildren, and his three sons in Army uniforms.

> *They all went to Vietnam, and then they passed that law that two or three brothers didn't have to fight in the same war. Ruby got in touch with their company commanders and things like that, and two of them could have come out of there if they'd of wanted to. And they got in touch with the boys, and they said, "To hell with it. We'll just all stay over here and fight." So they all stayed. One of my sons was wounded. A sniper got him in one side of the chest, and the bullet come out the other side. It just went under the skin and traveled all the way across his chest and come out another hole. It buggered up his chest pretty bad. But he was the only one that got wounded. And none of them was captured. They pulled their time over there, and they all come back.*
>
> *One of my boys, he thought he might want to mine coal, and I took him up here to Brookside and got him a job. He worked two weeks. He told me, he said, "I just don't believe I like coal mining. I don't like working down in a hole." And he went right back to Detroit. After two weeks. And that's the only one that's ever tried working in the mines. The other two sons, why, they just said they didn't want nothing to do with coal mining. I never would encourage them to go in a mines. I told them if they didn't want to, not to do it. On account of it being dangerous work.*

He lit another cigarette and blew the smoke across the room, toward the pictures of the boys in uniform. "Yeh, it's pretty quiet around here since my kids all growed up. About all we ever do is go to the grocery store, go to the union meeting and out on the picket line." He stopped, his eyes widening with a new thought. "Hey, Ruby," he said, "what we gonna do without that picket line? We're gonna have to take up something to occupy our time, ain't we? We might have to get out here and make us a hop-scotch ring. Or buy us a hooly-hoop."

They laughed that comfortable laugh of people who have laughed together for a long time. "Old man," she said, "you want to end up with broken bones?"

Later in the afternoon, the absence of a picket line began making Louie nervous. He got into his car and went off to find another miner to talk to.

Minnie Lunsford sat on the front porch of her white frame house with her husband, Willie, who was in the mines for forty-six years before he retired in 1965. Willie hunched forward in a white lawn chair, his elbows resting on his knees, staring silently across the yard. Minnie swayed gently to and fro on the white glider.

"Well, if they hadn't signed that paper last night, there'd a been a blood war," she said. "Those men had just sat out there in the rain and the snow too long, and they's just getting too tired. I won't say it's good that a man's dead, but it's better that one has given his life than maybe a whole lot more. We all have a time to die in different ways. I just feel that it was the almighty hand of God that this boy lost his life for the rest. It had to happen to get this strike settled before a whole lot more got killed. If it hadn't happened, there would be men going up and down this road amurdering. It comes to that."

Willie just grunted.

16 VICTORY & TEARS

SATURDAY threatened rain. The sliver of sky between the ridges seemed determined to spoil what miners said would be the biggest funeral Harlan County had ever seen. But by ten o'clock when the men began gathering at the Evarts Multipurpose Center, the clouds were breaking up, and sunlight began sliding down the steep mountainsides toward Clover Fork. The last day of August was going to be bright, after all.

The men filed into the hall quietly, as if into a church. In a few hours the large room would serve as a chapel, and the men seemed to have forgotten which piece of business was to come first.

Mickey Messer introduced a parade of union friends and big shots who had come to congratulate United Mine Workers Brookside Local Union 1974, as the men had been chartered, a number that none of the local's *present* members, at least, would forget. The speakers delivered the clichés that have resounded in meetings through the ages, whenever people have won something and there is a spotlight to be grabbed. The men applauded them politely, then got down to business. They would accept nothing on faith. They wanted to hear every detail of the contract. They asked questions about seniority, about the possibility that strike-related legal charges would be dropped, about initiation fees, about the coverage of the welfare fund. Mike Trbovich, Harry Patrick, Tom Pysell, and other union officials answered them patiently.

"What about that no-strike clause?" someone called.

Eastover and Duke had agreed that they would also sign the 1974 national contract as soon as it was negotiated, Pysell replied. In return, the Brookside men would not walk out again if the union's other 120,000 workers struck in November. "You've been on strike for thirteen months," he said, "and that's long enough for anybody. You've already got the 1974 contract in advance. If that's a no-strike clause, I wish we could get it into *every* contract."

It was moved and seconded that the contract be accepted. There was no dissenting vote. The men rose and filed forward to sign for their weekly strike benefit checks. As they walked out, the undertaker's people filed in, carrying huge armloads of flowers.

The organizers herded the miners together on the bank of Clover Fork for the official victory photograph. Behind a banner handpainted with the name of the new local, they stood, laughing and joking again, holding aloft their fingers in the victory V, holding the neatly printed signs they had carried in their "memorial" parade a week before:

HARLAN
COUNTY
IS
UMWA
COUNTRY

In early afternoon, the funeral crowd began to gather, quickly filling the seats in the meeting hall, filling windows and all the standing room there, overflowing into the lobby of the Multipurpose Center, then out onto the sidewalk and into the muddy parking lot. Men and women who arrived an hour before the service saw it was futile to try to get into the hall. They stood in small knots, talking. An NBC camera crew scurried about, filming the faces of the talkers and the cars and pickups still coming down the hill from Kentucky 38.

There was a brief flurry of excitement when Arnold Miller arrived, silver-haired and somber. NBC interviewed him. "This is just the beginning," he said. "The union's coming back to Harlan County." Miners and pensioners buttonholed him, slapping his shoulder, shaking his hand. He listened, then moved on into the hall.

Soon the funeral entourage arrived, the shiny hearse and black funeral cars out of place among the muddy, bumperstickered miners' cars. The young, long-haired pallbearers, one furiously chewing gum, some dressed in ill-fitting black suits and white cotton gloves, manhandled the bright blue casket out of the hearse and up the sidewalk.

While the crowd looked on silently and the TV camera whirred, the family followed, tightly grouped, walking slowly, young men holding the arms of Lawrence Jones's mother, who wept loudly and cried words incoherently into her white handkerchief. The widow Diane, looking too young to weep over anything more tragic than the loss of a football game, followed, tears streaming down her cheeks and dropping onto the frilly new dress of her infant daughter.

On the stage, the nattily dressed officials of the United Mine Workers of America sat beside the coatless, short-sleeved singers, the pianist, the two guitar-pickers, and the preachers, looking down on the blue casket. As the choir bellowed out the sad old mountain hymns, the keening of the women grew in intensity until it pierced through the music, its pain overwhelming the sounds of comfort.

Trbovich rose and read a brief eulogy, making no martyr of Lawrence, calling him merely a "fallen brother of the United Mine Workers of America."

The Reverend Henry Long, his shirt collar open, sweat glistening on his haggard face in the television lights, paced nervously up and down the stage as he preached, his fingers entwined, fluttering like small birds. "It seems like this funeral has done more to me than any other funeral," he said. "It's brought me to a low I can't explain."

He described his visit to the hospital after he heard of the shooting, and how his spirit sank when he saw the damage that had been done. "I thought, 'Lawrence, I can't do nothing for you now.' "

Two of Lawrence's young friends sang and picked their guitars, keeping their eyes above the casket and the weeping women. Then the Reverend Homer Jackson, whose hay Lawrence had hauled on the day he was shot, warned the people to remember that Lawrence had died young. "He was going about his business, and God reached out and brought him in. And he can reach out and get you, too."

The preachers were not in a mood to preach long, and soon the union officials were standing beside the casket and greeting the dead man's friends and neighbors as they filed past. Picket-line veterans who had been angry for so long now fought back tears, and women broke into incoherent wailing. Some stopped and sat until their shaking calmed.

Tom Pysell, who had explained the contract to the victorious miners only a few hours earlier, supported Lawrence's mother until the hundreds of mourners had passed. Then, as she approached the casket herself, she collapsed, and Pysell carried her out of the hall.

The procession moved slowly around the sharp curve where the Battle of Evarts was fought forty-three years before, and on down Kentucky 38 to the grassy little graveyard on the hillside, where the mourners laid Lawrence in the ground.

Far below him was the cool, dark mine where he had worked for such a little time.

17 LOUIE AGAIN

OUTSIDE the small kitchen window, the hummingbird moved among the marigolds. The dog Cricket was asleep on the sofa. Ruby was rocking in the big chair.

"Well," Louie said, "I guess I can clean up the old shotgun and let her rest awhile. She got all rusty, out there in the rain and the dew."

He yawned and stretched in the gloom. "I have something to look forward to now," he said. "Thirty-eight months straight work. Let's see, providing the Old Man Above sticks with me, and I don't get maimed up or sick or killed, I'd say I'm good for forty thousand dollars or more in the next thirty-eight months."

"Now, Louie, you hush," Ruby said. "You know better than to go talking like that."

"Aw, Ruby, wait a minute now. I said *providing* the Old Man Above stuck with me. *Providing* I don't get killed or sick or maimed up. Now I've got *conditions* in there. You heard them, didn't you?"

They laughed, then fell silent. Ruby got up and turned on the light.

The Brookside Strike Is Over

JUNIOR DEATON

The Brookside strike is over,
After many months of strife;
It's back to the pits for the miners,
And back to the kitchen for the wife.

At times it seemed to be lonely,
But at times we were very gay;
But we were assured when all was done,
The UMWA would be here to stay.

We didn't have the best of everything,
But we did have plenty to eat;
Although we had to sacrifice,
We didn't suffer defeat.

The women, bless their hearts, they're good;
They often came when needed;
They were dragged and jailed and criticized,
But they never were defeated.

Before the women entered the cause,
The company had lots of scabbing;
But after a few days of woman's ways,
There was nothing going on but blabbing.

We give lots of thanks to the media;
We think they were very concerned;
We hardly think they wasted their time,
But rewarded with what they have learned.

We think what they did was important,
In making our cause well known;
We wish them well in every respect,
Now that most of our troubles are gone.

We wish the company every success,
And we aim to try to please;
We know we can produce the coal they want,
And keep both parties at ease.

Our greatest thanks to our leaders,
That kept us from going astray;
You've gained our respect for what you've done;
Now may God be with us each day.

Let us pray for the family of the brother we lost;
Their hearts will remain full of sorrow;
We shall never forget the price they have paid,
In providing us a better tomorrow.

EPILOGUE

ON MAY 22, 1975, Billy Carroll Bruner took the witness stand in his own defense.

He had been on his way home on that fatal afternoon, when one of Lawrence Jones's friends hailed him, he said. He noticed that the men were armed. He parked his truck, picked up his shotgun, and walked down to the road.

"I was scared to death when I saw what I got into. I don't know who fired first. I was surrounded, and there was shooting all over the place. . . . I fired but not *at* anybody. I was trying to get them down, so I could get behind my truck. Shots were coming from all directions. . . . I didn't know Lawrence was shot. I was pinned down. I couldn't see nothing. . . . I had bought the shotgun two or three weeks before to protect myself."

"To protect yourself against what? Bears?" the prosecutor asked.

"People," Bruner replied. "We were having labor trouble at Highsplint. They had went on the picket line. They had set tacks. . . ."

"You weren't the only Highsplint employee going to work, were you?"

"I wasn't the only one being harassed, either."

After nearly two hours of deliberation, the jurors declared themselves unable to agree on a verdict. The judge ordered them to try again. They pronounced Bruner innocent.

None of the Brookside men or their wives attended the trial.

Another trial never happened. While Lawrence Jones lay dying, a group of Duke Power Company stockholders, armed with information supplied by UMWA researchers, threatened to sue the president and directors of their company for gross mismanagement. Had such a suit been filed, and had it succeeded, the corporate officers might have been held personally liable to the company they managed. Some believe it was fear of this suit—and not remorse for the death of a miner or dread of a "blood war"—that sent Norman Yarborough and Carl Horn winging to Washington to end the strike.

U.M.W. OF A.
ON STRIKE
Brookside
PUT YOUR IN U.MWA
OR GET YOUR OUT
I PREFER A.UNION MAN
FOR A FRIEND. HOW ABOUT
JOINING THE U.M.W.A. AND
BE A GOOD FRIEND
SACRIFICE.A.LITTLE.TODAY
TO.HAVE.A.GOOD.SAFETY
PROGRAM.TOMORROW
BROOKSIDE.WILL.NOT.BE
SCABED. ALL OF US UNION.MEN
OUR.WIFES CHILDREN.LABOR
BOARD. RETIREES. C.B.S. AND.THE
WHOLE. U.S.A. HAS.A.BIG
ON.EASTOVER
AND. WE.ARE.NOT. ASLEEP
I. DID.NOT. HAFT.TO.USE
SOUTHERN.LABOR.TODAY
CHANCES.ARE.I. WONT. HAFT
TO.USE.IT. TOMORROW
Louie Stacy

The coal camp at Brookside

The Brookside tipple

The window of Deaton's Grocery

The Highsplint picket line, July 1974

The Highsplint picket line

STATE
POLICE

Tub Rainey's truck in front of the Eastover office

BATH HOUSE
Coca-Cola
UCKIES

DEATON'S GROC.
KERN'S
KERN'S
RC
DEATON GRO.

I'M A COAL
MINERS
SON
RID YOUR HOME
OF
ROACHES, ANTS
AND
WATERBUGS
WITH
CAPITOL
PUNISHMENT
ONLY $1.49
UMWA
Candyland
"REALLY FRESH!"

Union men in Deaton's Grocery. Tub Rainey is at right.

BIG BLACK MOUNTAIN
Lonnie's Speedway
UMWA
HERE TO STAY
Harlan County Is
UMWA Country
THIS BUSINESS SUPPORTS THE

Junior Deaton

UMWA rally, July 21, 1974

Arnold Miller talking with a retired miner at the UMWA rally

UMWA:
Safety
or Else
UMWA

William Crusenberry, a retired miner, with
his son at the UMWA rally

The Brookside picket line, July 1974

The UMWA march through Harlan, August 22, 1974

Darrell Deaton with his wife and daughter

Wilma Osborne, wife of a miner, with two of her children

Striker Bill Broughton and his family in their house
at the Brookside coal camp

Tub Rainey and three of his seven children on the back porch of their house at Brookside

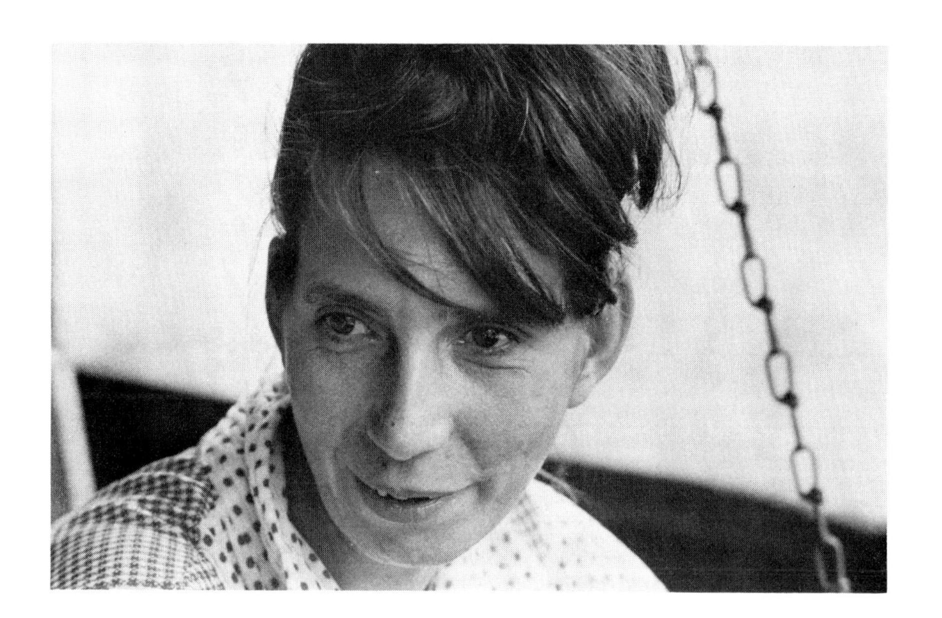

Nannie Rainey

The Reverend Otis "Tag" King with his great-grandson

BILLY G. WILLIAMS
FOR
SHERIFF
YOUR LOCAL WELL DRILLER

Coca-Cola
STOP
BLACK
LUNG
MURDER
Don't Scab For Duke
Support UMWA Brookside Miners

Don't Scab For Duke
Support UMWA Brookside Miners
SUPPORT
JOIN UMWA
STP
an
after-
thirst
THE PEOPLE'S
WORTHINGTON

Willie Lunsford, a miner for forty-six years

Minnie Lunsford

Minard Turner, who was shot on the Highsplint picket line in July 1974

"Tag" King

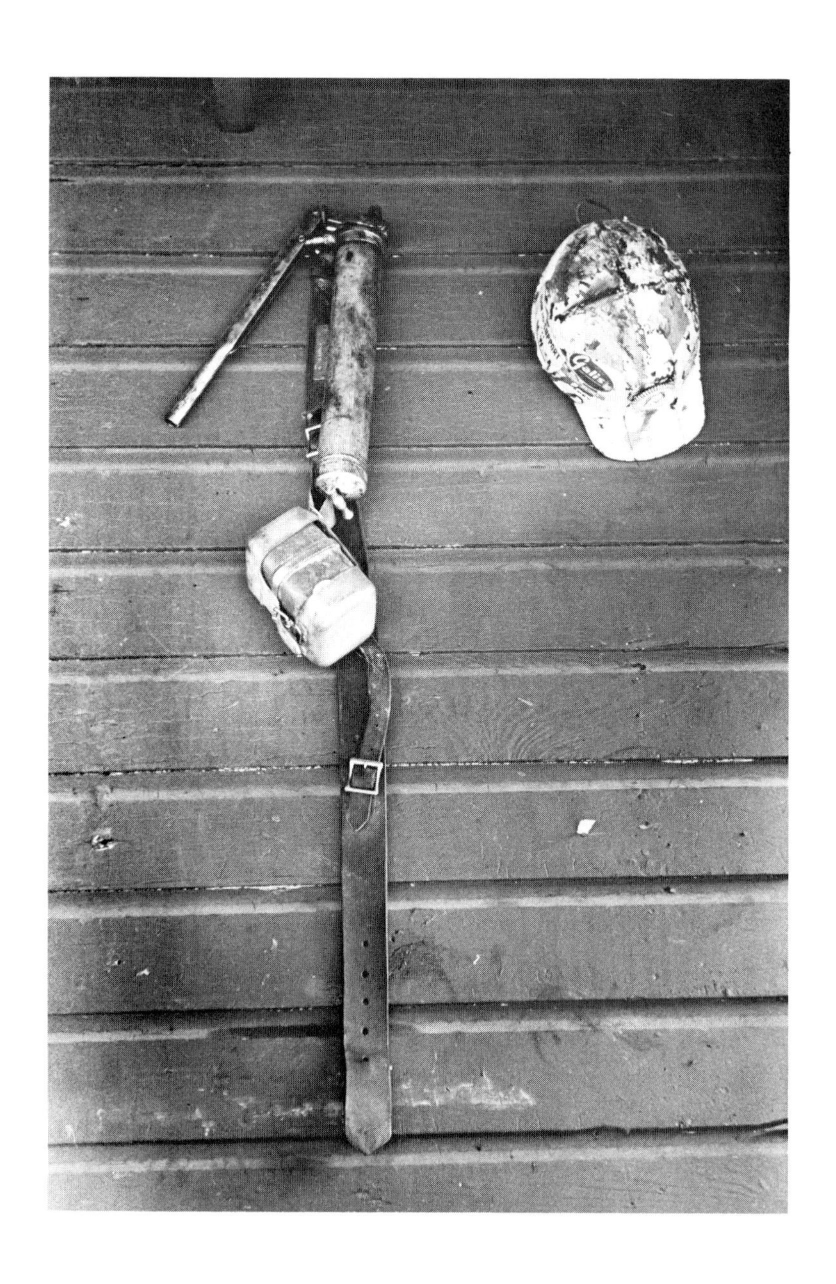

Tub Rainey's hard-shell hat, miner's belt, and self-rescuer

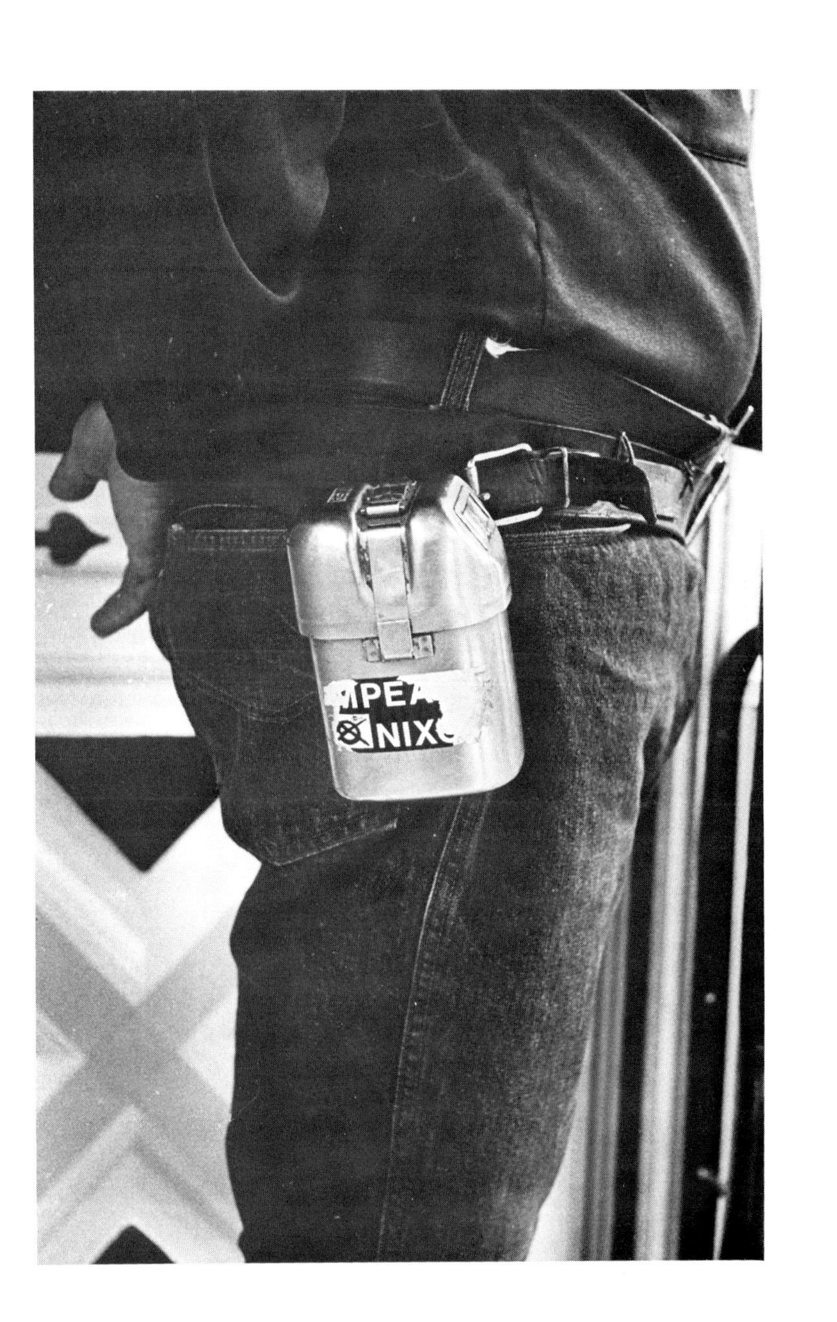
MPEA
NIX

Louie Stacy being interviewed by a CBS television crew

UMWA
HERE TO STAY
Harlan County Is
UMWA Country
THIS BUSINESS SUPPORTS THE
UMWA IN ITS EFFORT TO
ORGANIZE HARLAN COUNTY
UNITED STATES DEPARTMENT OF AGRICULTURE
we accept...
FOOD COUPONS

On top of the Stacys' TV

Louie with his shotgun, Ruby with her "switch," a length of broomstick wrapped with tape and adorned with a UMWA bumper sticker

Lawrence Jones's funeral

Outside, looking in at the funeral

UMWA L.U.1974, BROOKSIDE KY.
HARLAN COUNTY IS UMWA COUNTRY
UMWA HERE TO STAY Harlan County Is UMWA Country
HARLAN COUNTY IS UMWA COUNTRY
HARLAN COUNTY IS UMWA COUNTRY
UMWA HERE TO STAY Harlan County Is UMWA Country

Back to work. Louie waiting for a ride.